Metal–Crucible Interactions

Metal–Crucible Interactions suggests solutions to a major challenge in high-temperature materials processing. It offers a holistic presentation of the current knowledge of metal–crucible interactions in a compact volume so that readers can take informed decisions on materials selection. Presenting practical information, this book:

- Provides an extensive summary on the compatibility a huge variety of metal–container combinations, assembles information about all known significant interactions, and evaluates how they are managed
- Explains the underlying reasons for the occurrence and extent of incompatibility between metals and containment and presents some possible solutions
- Outlines analytical experimental techniques to quantify compatibility/incompatibility
- Covers issues and resolution in interrelated solid–solid, solid–liquid, solid–gas and solid–liquid–gas processes determining compatibility
- Discusses all the metals - ferrous, common non ferrous, reactive and refractory metals, rare earths, and the important alloys as well as compounds and special compositions that tide over or remain prone to degradation due to compatibility issues
- Highlights the value of addressing all interrelated issues in arriving at durable solutions to compatibility challenges

Aimed at readers in industries dealing with materials processing at high temperatures, research scientists and engineers, and graduate students, this book addresses a topic vital to stimulating immediate and long-term research and development benefits, in ways not previously covered in other books.

Metal–Crucible Interactions

Nagaiyar Krishnamurthy

CRC Press
Taylor & Francis Group
Boca Raton London New York

CRC Press is an imprint of the
Taylor & Francis Group, an **informa** business

First edition published 2023
by CRC Press
6000 Broken Sound Parkway NW, Suite 300, Boca Raton, FL 33487-2742

and by CRC Press
4 Park Square, Milton Park, Abingdon, Oxon, OX14 4RN

CRC Press is an imprint of Taylor & Francis Group, LLC

© 2023 Taylor & Francis Group, LLC

ISBN: 978-0-367-34809-0 (hbk)
ISBN: 978-1-032-30787-9 (pbk)
ISBN: 978-0-429-34556-2 (ebk)

DOI: 10.1201/9780429345562

Typeset in Times
by Deanta Global Publishing Services, Chennai, India

Sri Krishnarpanam

Contents

Preface

Metal–crucible interaction is a nearly universal problem, which every researcher and materials processor faces, particularly when dealing with metals and materials near, at or above their fusion temperatures. It is usually a deleterious interaction between the container and the melt. It is not entirely preventable but invariably managed to levels that are considered harmless in the given context. This has been the general situation from the beginning of metallurgy thousands of years ago and is still the case today. The know-how and the resources to contain molten materials have determined the manner and extent to which materials processing progressed in various regions of the world and the range of materials to which various civilizations had access. In modern times, too, this situation prevails somewhat implicitly.

Every beginning researcher in metallurgy wants and searches for a ready-to-use table that lists various melts and the crucible materials that can be used with them. Some are available, mostly in the manuals that come with expensive high-temperature equipment, and some from colleagues and associates who have created their own cheat sheets. While such collections are attractive and comforting, their use is very limited, because each situation of metal–crucible interaction is unique. A sneak preview into the real world of metal–crucible interactions should bring out the challenges in all their glory.

First of all, it is necessary to abandon the common perception that a crucible is nothing but a cup-shaped vessel that is used to hold materials at high temperatures. A crucible is much more than that. More inclusively, a crucible can be considered as the material surface at work in contact with the melt and related fluids during material processing; the work here being to contain the melt, resist and survive its attack, and withstand the concurrent thermal and mechanical stresses the process entails, and to do all these without causing any degradation of the melt. The crucible also needs to maintain its integrity and functionality for durations that are useful in practice. In the field of common metals, both ferrous and common non-ferrous, the crucibles are usually refractory ceramic materials. In the case of less common metals, the crucibles are more specialized and are often made of refractory ceramic compounds, refractory metals and also some special alloy compositions.

Crucible materials include metals and alloys, oxide minerals, pure oxide compositions, and compounds like metal borides, carbides, nitrides, silicides, phosphides, sulphides and even fluorides. Carbon is an exceedingly useful crucible material, mostly as graphite and also as vitreous carbon. Many of the properties of these materials are available not only when they are used in monolithic form but also when they are used as linings or coatings on other materials. An impressively versatile range of applications is fulfilled by the class known as water-cooled crucibles, which generate a layer of solidified material (or skull) on the working surface and thus obviate any chance of material contamination. Refractory materials for crucibles are broadly grouped into siliceous, aluminous, aluminosilicates, magnesite, dolomite, chrome ore, and a miscellany of materials known as special refractories. Each of these refractory types has its niche application areas. Every crucible material is subject to attack by the melt it is holding. Melt penetration into the crucible and reaction of the melt with the refractory occur. Generally, refractories have the ability to withstand alteration by penetration, contamination and/or reaction in service and still function as reliable engineering structures. Many factors control this interaction, and some of these are manageable to a certain extent.

Metal–crucible interaction operates by chemical interaction between the material of the crucible and the melt, augmented by concurrent physical, mechanical and thermal processes. Chemical interaction, in turn, generates products that, because of their own properties, occasionally end up protecting the crucible but usually result in its degradation. The key process, therefore, is the chemical reaction, and chemical reactions are governed by thermodynamics.

Even though the statement that at sufficiently high temperatures, everything reacts with everything else, cannot be disputed, it may be tempered with many ifs and buts. These conditions, which are extremely important in real life situations, are physical, structural, mechanical and thermal. A crucible material should possess, in addition to high thermodynamic (chemical) stability, high melting point, good strength, appropriate thermal expansion and thermal conductivity and hence, thermal shock resistance. Some of these properties are interrelated and can be tailored by microstructure and porosity control to ensure that the chemical stability of the crucible material is fully utilized in actual applications.

To interact, the melt and crucible have to be in close contact. This aspect is decided by the wetting characteristics of the melt–crucible combination. The contact angle is a measure of wetting and is determined by the interfacial tensions of the pairs: liquid–solid, liquid–vapour and solid–vapour. The contact angle or wettability is ultimately determined by the real-time chemical composition of the melt–solid interface. Impurities in the melt or crucible play a role, and this is the basis of anti-wetting agents. Anti-wetting additives act by controlling the interface chemistry and raising the contact angle. They have been used extensively with aluminium melts because of the temperatures involved. Nearly all anti-wetting additives decompose at higher temperatures, and then, the protection will be lost (usually beyond 1000–1200 °C). For this reason, porosity control may turn out to be the more reliable handle to obviate melt intrusion. However, modifying porosity may have other unintended consequences.

Corrosive attack on the refractory of the crucible is often due to the combined action of molten metal, slag and also process gas. The most significant and life-limiting attack is by the slag and also by metal that acquires the characteristics of slag. The attack proceeds in two ways. Direct dissolution is controlled by the rate of chemical reaction at the slag–refractory interface. Indirect attack is controlled by the rate of diffusive transport through the slag or through the newly formed solid phase. Significantly, penetration and thus, corrosion are determined by local slag composition and also refractory composition and structure. These are also usable handles. The melt properties, particularly viscosity, are an important factor in the corrosion process. Many attack prevention strategies rely on controlling this property by temperature and/or composition manipulations.

Throughout the history of metals processing, metal makers have faced the challenge of finding suitable crucibles for smelting as well as post-reduction melting and casting operations. The challenges have been faced and circumvented with ingenuity, directed experimentation and sometimes plain luck, and there is now available a large and valuable body of knowledge on what works and sometimes, also the underlying reasons for the success. The situation as regards some of the major metal–crucible systems is dealt with not only to look back at the road travelled but also to prepare for the road that lies ahead.

The purpose of this book is also to consolidate in one place all the different metal–crucible combinations that have been tested and found to work, and in the process, look for the common underlying reasons that make such compatibility a reality. Materials interact or fail to interact because of certain well-defined reasons. The reasons concern thermodynamics, on the one hand, and transport processes at the interface, on the other. How these factors play out in each individual case is unique, but as is generally stated, there is method in the madness. These methods are rather implicit and are best revealed by describing the working combination. It was therefore considered best to select certain major metal–crucible systems and look at the key processes involving the crucibles from the point of view of the crucibles. The systems considered in this book are iron and steel, aluminium, copper, reactive rare metals like uranium, niobium and vanadium, rare earths, and last, and in considerable detail, titanium and its alloys.

Each of the few metal–crucible interactions selected for specific coverage in this book is representative of the type of problems faced by many more metal–crucible systems. The drivers for interactions and the remedial actions that can circumvent or offset the damages are described. The objective is to relate what is known about metal–crucible interactions at the basic level to the interactions playing out in real processes. This will help the task ahead as regards managing new metal–crucible combinations and suitable approaches that would improve the chances of success. Most of the issues concerning any metal–crucible pair can usually be resolved by analogy to those covered specifically here.

Systems with melts and containers in contact at high temperatures are encountered in two major areas of technology. One is in process metallurgy and post-reduction processing of materials, and the other is in liquid metal/alloy cooling loops. Both are important and have been attracting considerable research effort all over the world. There are, however, major differences in the problems faced in each of these areas and the way solutions to the problems are developed. They have more differences than commonalities. To maintain focus, this book covers essentially the processing-oriented issues of metal–crucible interactions.

I began to mull over the various possibilities of creating a monograph dealing principally with metal–crucible interactions soon after my book on rare earths was completed in 2015. The break came in July 2017 when Florence Kizza, Editor, Engineering & Environmental Sciences, Taylor & Francis enquired whether I would be willing to author a title in their 'Focus' series. I was, and prepared a proposal for a book on 'Metal–Crucible Interactions'. My mentor at T&F, Allison Shatkin, received the proposal, and the process began. I was obviously underestimating the matter that should be covered under this title when I originally aimed at fitting everything into 150 pages, the typical size of a 'focus' title. The highly experienced international review panel who took a look at my proposal concluded otherwise and suggested a bigger book with at least 250 pages. So, here it is.

During the preparation of the manuscript, which took considerably more time than I intended, I was fortunate to receive cooperation and support from both the publisher and my family. I have been constantly encouraged and guided all along by Allison Shatkin. Gabrielle Vernachio has been extremely helpful. She was in constant communication with me, answering my numerous queries concerning the presentation of text and figures. Florence Kizza started the project, and Camilla Michael, Ryan Farrar, Iris Fahrer and Lara S. Loes have at various stages provided support My sincere thanks to all of them.

All through the course of this work I received wonderful cooperation and support at home from my wife Kusuma and daughter Kavita. My son Kumar had always urged me think of creating monographs that I would have liked to possess myself during my formal working years. I would like to attribute all good things in the book to these wonderful people. If any, errors and omissions, are entirely my doing. If brought to my notice, I shall remedy them in future editions.

Nagaiyar Krishnamurthy

Acknowledgements

The role of an author in creating a monograph is aptly described by the phrase 'standing on the shoulders of giants', a metaphor coined by the twelfth-century French philosopher Bernard of Chartres and made famous by none other than Sir Isaac Newton in his letter to Robert Hooke in 1675. As an author, the extent of support I received from people and institutions has been truly humbling.

Nearly all scientists and engineers whose work involves the processing of materials at high temperatures have been confronted with issues of interaction between melts and the containers. In their publications, they often highlight the issues and also indicate if the interactions were circumvented or lived with. Original research papers, reviews and compilations, and technical reports from government laboratories, over the years, contain a wealth of information on this matter. Then there are books and conference proceedings that both allude to or directly deal with technical details on the interactions and also their ramifications in the overall processing. In putting together the present volume, I have strongly relied on these sources of information that have been made accessible to anyone wanting to read or use them by the wonderful publishing houses and organizations. I have greatly benefitted from and gratefully acknowledge such publications brought out by Elsevier, Springer, Wiley, Taylor & Francis, Maney Publishing, Routledge, Vulkan-Verlag GmbH, The AISE Steel Foundation, EDP Sciences, American Ceramic Society, American Chemical Society, Quartz Business Media, The Minerals, Metals & Materials Society, American Society for Metals, as well as the Institutions of the U.S. Government and the Government of India. Wikipedia and the Encyclopaedia Britannica often provided the linkage when the terrain got tough. The open access system in technical publications is truly an exalted level in information dissemination, and every author will remain deeply grateful to all those responsible for its coming into being.

In addition to all those wonderful people from Taylor & Francis mentioned in the Preface, I gratefully acknowledge the contributions of Ms. Jayanthi Chander, Project Manager, and her team at Deanta Global Publishing Services, Chennai, in bringing out the book in its final form.

Nagaiyar Krishnamurthy

Author Biography

Dr. Nagaiyar Krishnamurthy was affiliated with the Materials Group, Bhabha Atomic Research Centre, Mumbai, India for nearly four decades, conducting and later also formulating and guiding research in the extraction and processing of rare earths, reactive and refractory metals, and special less common materials. He earned his BSc degree at the University of Madras in 1974, his MSc in 1980, and his PhD in 1992, both at the University of Bombay. His PhD dissertation was on the pyrometallurgy of Group V refractory metals and their alloys.

He has been professor at the Homi Bhabha National Institute and has guided many students to complete their PhD. Apart from regular classroom teaching of postgraduate engineering and science students at the Bhabha Atomic Research Centre Training School, Professor Krishnamurthy is a popular faculty member in various quality improvement programmes for college and university teachers. He has been active as a lecturer in popular science topics connected with materials and has lectured to audiences ranging from high school and college students to members of staff in universities and national institutions.

In addition to more than 100 original research papers published in peer-reviewed international journals, Dr. Krishnamurthy has co-authored two editions of *Extractive Metallurgy of Rare Earths*, published by CRC in 2004 and 2016. Earlier, he co-authored the book *Extractive Metallurgy of Vanadium*, published by Elsevier in 1991, and a monograph, *Binary Phase Diagrams of Tantalum*, published in 1996 by the Indian Institute of Metals and American Society for Metals. He also served as a key reader for the journal *Metallurgical and Materials Transactions*.

1 Metal–Crucible Interactions

Evolution of Crucibles

1.1 INTRODUCTION

The earliest historical reference to metals dates back to 10,000 BCE, which is approximately the age of the oldest metal artefact found anywhere in the world. It is made of copper. Between then and the beginning of the Common Era, 10 more elements, of which 8 were metals, were added to the list of elements discovered. It is amazing that so many great civilizations were sustained with such a small number of metals. Furthermore, by the beginning of the industrial revolution (1720 CE), only three more elements had been discovered. These were arsenic, antimony and phosphorus. The pace picked up by about 1720 CE, and one by one, all the remaining 72 naturally occurring elements, which include 59 metals, were discovered, and many of them were put into use. The most prevalent processing method was melting and casting. The facilities for melting a metal often determined whether the metal could be used and also, how widespread its use would be. Suitable crucibles were needed for melting metals, and the availability of a crucible was often the decisive factor in metal processing and use.

The concept of the crucible evolved from the necessity to confine materials within a physical boundary when processing them, especially by raising or lowering their temperature. The process may be melting and casting to consolidate the smaller pieces into a solid mass, a chemical reaction by bringing together the selected reagents, or the controlled cooling of a melt to develop some properties in the solid.

The evolution of crucibles is, therefore, linked to the types and amounts of materials they needed to hold, the manner in which heat was supplied to or withdrawn from them, and the availability of substances out of which the crucibles could be made. A large variety of materials, both non-metals and metals, have been found suitable to function as crucibles. Chemical, physical and mechanical interactions always occur between the crucibles and the materials they hold. The extent of these interactions determines the quality of the products obtained at the end of processing, the integrity and longevity of the crucible itself, and above all, the economics of the process. Seldom were there perfect crucibles, but workable selections have always been made.

The range of sizes and shapes of crucibles used in materials production and processing is vast. This can be fully understood if we consider a more inclusive definition of a crucible as a confining solid boundary within which processes involving

DOI: 10.1201/9780429345562-1

chemical reactions and transport of heat and mass occur to yield a physically sepa-rable metal or an alloy in a certain state of purity and crystalline perfection. It then becomes possible to consider any metallurgical reactor as a crucible.

Crucibles have been used in one form or other for thousands of years. The mate-rials and forms of crucibles have evolved in response to the needs of processing more and newer materials throughout history. The elements that occur in nature have always existed, and people in every time period would have realized the existence of some of them. But, the ability to identify, access and use the metals was slow to evolve. The pace of this evolution has been determined by the availability of cru-cibles, among other things. The choice of crucible was a challenge at the very begin-ning of metals processing thousands of years ago and surprisingly, still remains a challenge today. Only the metals that pose the challenge have changed.

1.2   HISTORY OF METALS PROCESSING

The three-age system of describing human history by dividing it into the Stone Age (ca. 2,500,000–3200 BCE), the Bronze Age (ca. 3200–1200 BCE) and the Iron Age (ca. 1200 BCE–100 CE) was propounded by Christian Jürgensen Thomsen, the Danish antiquarian, in 1836. He divided prehistory into discrete periods based on archaeological evidence of the prevalence or the exclusive or predominant use of certain metals.

1.2.1   STONE AGE

The evolution of humankind has been closely linked with the use of tools. The vari-ety and types of tools possible, in turn, depend on the range of materials available. Stone tools, many types of them but nevertheless only stone tools, both naturally existing and improvised, were used in the incredibly long Palaeolithic and Neolithic eras (ca. 2,500,000–3200 BCE).

The Palaeolithic period began 2.5 million years ago, with the first evidence of human technology (stone tools), and ended with some stabilization in human societ-ies marked by the invention of agriculture and the domestication of animals. The Neolithic period, which corresponds to the first farming societies, extended from 6000 to 3200 BCE. Earthenware pottery and some shaped and polished stone tools, such as axes, began to be used. Metals, especially copper, gold, silver, lead and tin, had already made their appearance sporadically in the form of ornaments, charms and small tools from around 9500 BCE.

1.2.2   NATIVE METALS: COPPER, GOLD, SILVER

That the eras of civilization are named after metals highlights the exceptional influ-ence of metals on human life from the start. Interestingly, the knowledge that there are metals and that they are useful was revealed to our ancestors by nature present-ing some of the most important and useful metals in native forms. A metal in the un-combined or elemental state in nature is considered to be occurring in the native form. Native gold was relatively common and occurred as small flakes separated from the original deposits by weathering and transportation. Native silver and copper were less common. Copper in the native form was invariably found in small amounts but in more widely distributed locations (Bunch and Hellemans 2004). Native iron was found in locations where meteorites had deposited it on the surface of the earth in large masses of iron or as an alloy of iron and nickel. No occurrence of native iron could be linked to processes in the earth's crust.

Nearly10,000 years ago, copper was already familiar to humans. The earliest known metallic artefacts are small pendants, beads and pins of copper from the period between 9000 and 7000 BCE. Native copper was shaped by cold hammering in the period from 7200 to 6600 BCE. Along with beads of native copper, beads of fused lead, dated to 6500 BCE, have also been found in the ancient Anatolian mound of Catal Hüyük (Darling 1990).

For about five millennia, copper was probably the only metal known to man and fulfilled all the applications that required a metal. Although lead was also known at a very early date, copper was the first metal to be put to use. In the very early stages, humans were first tasked with the processing of metals rather than preparing them. Native copper was hammered directly into small articles of jewellery or good luck charms.

1.2.3 PRIMITIVE CRUCIBLES

In North America, implements and weapons were produced from native copper by cold forging with frequent annealing at temperatures below 800°C (Tylecote 1992). Interestingly, those craftsmen knew, even in 4000BCE, that copper could be hardened by deformation and softened by annealing, but they could not work out a means of melting copper. Indications of cast copper ingots were found much later in these locations. Excavations at the copper smelting site at Batan Grande in Peru have revealed a row of smelting furnaces. These furnaces were blown by blowpipes rather than bellows.

Between the seventh and fourth millennia BCE, most of the larger copper artefacts produced in the Middle East have a microstructure characteristic of crystals grown from the melt. In addition to hammer and anvil, the crucible became essential processing equipment in the earliest days of metallurgy.

The burning of charcoal for heating and intensifying the heat using draught, either natural or forced, were techniques known to humans well before any processing of metal was attempted. Various configurations were developed to implement these techniques. When native copper was first melted intentionally, it would have been heated from above by charcoal fire. Beneath the fuel heap, the metal was placed in a clay-lined saucer-shaped depression in the ground (Figure 1.1). The molten metal would have run together, collecting and forming a convex ingot. This is probably close to the first ever crucible and furnace arrangement used for processing any metal. Provision of a forced draught, through a tuyere connected to bellows (Figure 1.2), and also a chimney would have completed the rudimentary furnace arrangement to attain a higher temperature. The arrangement is exemplified in the Minoan

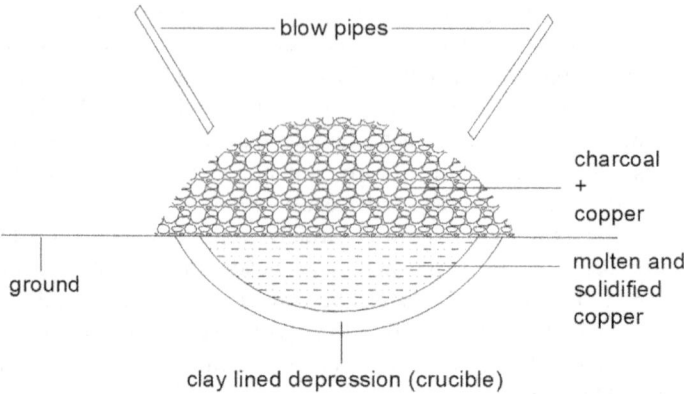

FIGURE 1.1   Primitive metallurgical furnace – crucible and blowpipe.

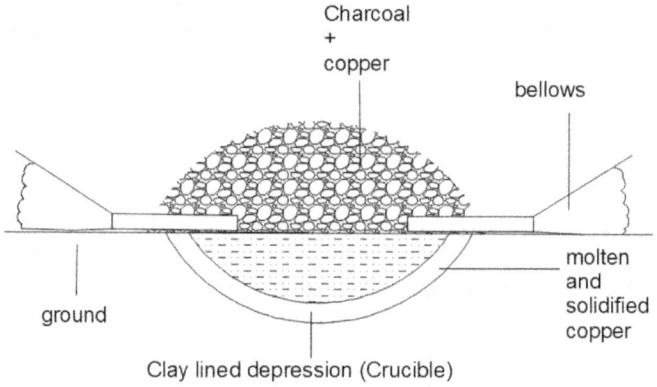

FIGURE 1.2   Primitive metallurgical furnace – crucible and tuyere.

furnace for copper smelting and casting shown schematically in Figure 1.3. The Minoan civilization flourished from ca. 3000 BCE to ca. 1450 BCE on the island of Crete and other Aegean Islands (Darling 1990).

To produce objects such as flat axes or mace heads, which were cast directly to size, crucible furnaces were used. These furnaces (Figure 1.4) had a vertical cylindrical clay shaft. Air entered freely at the lower end, providing the required draught. A hemispherical clay crucible, about 10 cm in diameter, was supported about halfway up the shaft by a bed of charcoal packed into the base of the furnace (Darling 1990).

Gold was known in Egypt well before the fifth millennium BCE. Gold was refined by fire created by the charcoal–blow pipe combination. Charcoal served as the crucible. Gold (m.p. 1064°C) was melted and held molten so that all other metals present, except silver, were gradually oxidized and separated.

Thousands of years later, after it was discovered in native form, copper was also the first metal to be smelted from an ore, by ca. 3500 BCE in Egypt and Mesopotamia.

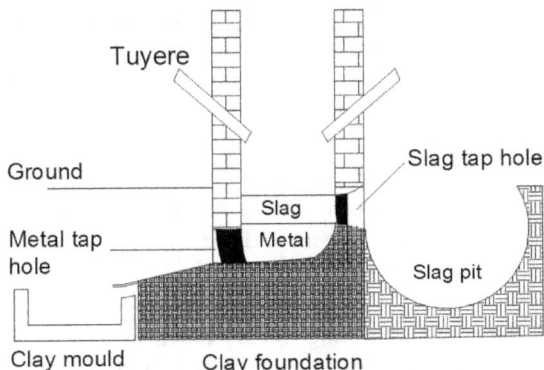

FIGURE 1.3 Minoan furnace for copper (From Tylecote, R.F. 1992).

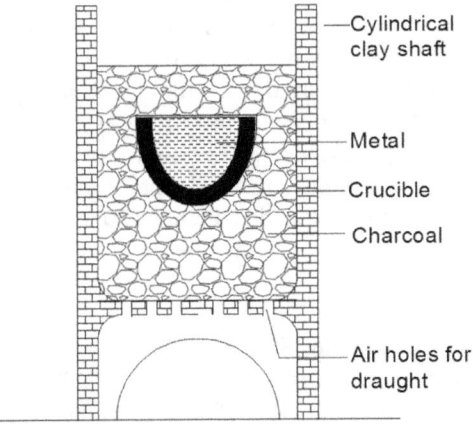

FIGURE 1.4 Chalcolithic crucible furnace (From Darling, A.S. 1990).

The list of firsts associated with copper is rather impressive. It was the first metal (native copper) to be cast into a shape in a mould, ca. 4000 BCE, i.e. centuries before it was smelted. Copper was also the first metal to be alloyed on purpose with another metal, tin, to make bronze, ca. 3500 BCE.

1.2.4 Chalcolithic Age

The beginning of the Copper Age or Chalcolithic Age is marked by the emergence of smelting methods to extract copper from its ores. The ores were bright green (malachite: copper carbonate hydroxide), blue (azurite) and metallic looking sulphide ores. It is surmised that malachite, used in Egypt as a green eye pigment, accidentally found its way to a wood or charcoal fire and was reduced to metallic copper. Once it was observed that fire could transform certain earthy and stony materials into metals, rocks or stones that were heavy or strongly coloured would have been thrown into or treated in fire to find out if they would yield metal. Apparently, many did, and pyrometallurgy was born in the most elementary form.

Archaeological finds in the 21st century have revised the probable first date of copper smelting to 500 years earlier. Copper smelting was probably practised from 5000 BCE in present-day Serbia. The fire obtained by burning the most common combustible materials – wood and charcoal – is not only a source of heat but also provides a remarkably gradable (tunable) reducing environment to the materials thrown into it. The combination of heat and reducing environment readily yielded certain metals, and these are the ones that were produced and so extensively used in early history.

Crucibles used for copper smelting during the Chalcolithic period were wide, shallow vessels made from the local clay, also used for domestic pottery, with melting temperatures around 1100°C. The charge was heated from the top using blow pipes (Rehren 2003). Thornton and Rehren (2009) also reported on an unusual metalworking crucible from the prehistoric site (dated to the middle of the fourth millennium BCE) of Tepe Hissar in northeast Iran. The crucible was made of steatite ($Mg_3(Si_4O_{10})(OH)_2$). It was apparently used for the melting and alloying of arsenical copper with lead. Steatite was also used in the middle Minoan period (ca. 2000 BCE) as mould material for casting copper (Tylecote 1992). Steatite is easily carved. In those early days in history, people had the knowledge of refractory minerals and the skill to selectively incorporate these materials into crucible making depending on process need (Darling 1990; Bunch and Hellemans 2004).

1.2.5 BRONZE AGE

The Copper Age, which could have begun as early as 5000 BCE, ended by 3200 BCE with the rising popularity of bronze. In the nearly 2000-year period from 3200 to 1200 BCE, the use of bronze (an alloy of seven parts of copper to one part of tin) was so widespread as the principal material in the manufacturing of weapons and implements that the period is given the name Bronze Age. Bronze is not only harder, stronger and more corrosion resistant than copper but also melts at a lower temperature (950°C). At the normal melting point of copper (1085°C), bronze is far more fluid, a property very conducive to casting intricate shapes accurately. The amount of bronze used in tools was dwarfed by the quantities used for making weapons. Tin ore does not normally occur with or near copper ores.

1.2.6 IRON AGE

The Iron Age, the successor to the Bronze Age, extends from 1200 BCE to the end of the first century CE. The Roman civilization was largely concurrent with the Iron Age and existed for 12 centuries in Italy, from the eighth century BCE to the fifth century CE.

The earliest iron samples are all of meteoric origin and are actually iron–nickel alloys. Iron was probably known for centuries in the Middle East before people living in present-day Armenia chanced upon a technique to enhance its properties.

They discovered that reheating and hammering iron produced a material greatly superior to copper or bronze. The discovery had military consequences.

1.2.7 MIDDLE AGES

The Middle Ages lasted nearly 1000 years. They began in 476 CE, at the end of the reign of Romulus Augustule, the last Roman emperor of the Occident, and ended in 1453 with the taking of Constantinople by the Turks. The progress of metalmaking during the Middle Ages was marked by technology diffusion rather than new inventions. Technologies fairly established in the East were becoming known and adopted in Europe. The route was through the Middle East. One was the making of steel for edge weapons, particularly swords. The production of cast iron and zinc metal was also becoming more familiar in the West during this period. Sporadic attempts were also made to build and use blast furnaces for producing molten iron and casting it. The Catalan forge was perfected for ironmaking by around the eighth century in Spain.

1.2.8 POST-MEDIEVAL PERIOD

The post-medieval period in metallurgy refers to the time period from 1500 CE to the beginning of the industrial revolution, i.e. 1750 CE. At the beginning of this period, the famous refractory crucible from Hesse, a region in Germany, became available. At about the same time, another specialized crucible was also made in Southern Germany. It was made of graphite. These had a very similar design to the triangular crucibles from Hesse. Reverberatory furnaces were developed, first for melting and eventually for smelting. There was a gradual increase in the size of blast furnaces, with more mechanization of air blowing by using water wheels.

1.2.9 INDUSTRIAL REVOLUTION AND LATER

In the period of the industrial revolution (from 1750 CE onwards), there were certain major changes in the way metals were prepared and processed. The steam engine, arguably one of the greatest inventions of the industrial revolution, added new power and reliability to the blowers and made the blast furnace, as it is known now, possible. Initially, it was the Newcomen engine and later, in 1775, the much improved steam engine of James Watt. The next was the Bessemer process for mass production of steel from molten pig iron. The process as introduced in 1856 had issues, which were eventually resolved. The removal of impurities from the iron via oxidation as air is blown through the molten metal turned out to be a more general technique amenable to much improvisation and modification for wider applicability. The invention of the electric battery in 1800 by Alessandro Volta led to the eventual development of the field of electrochemistry. Humphrey Davy used electricity to reduce many of the alkali and alkaline earth metals, Faraday came up with the laws of electrolysis, and Hall and Héroult revolutionized metals extraction by inventing the process named after them for electrolytically reducing alumina to aluminium. Amongst all these, a German inventor, Joseph von Fraunhofer, invented the spectrometer in 1814, and this became a new tool to discover more metals. The discovery of metals by this and other classical methods continued, and by 1939, all the naturally occurring metals, as well as other elements, had been discovered. The majority of these metals, however, did not appear to have any applications, and there was no incentive either to investigate them in detail or to produce them in good quantities. But, that lasted only until World War II.

1.2.10   WORLD WAR II AND LATER

The war and the concurrent development of atomic energy for military and civil purposes changed the metals scenario totally. Many metals that had been discovered and all but forgotten became relevant and important. The eventual advances in electronics and aerospace technologies added more metals to the list of materials that must be developed urgently. The period between 1940 and 1970 was a watershed moment for metals and materials development. The metals originally produced and used for centuries came to be called common metals, and the newly important metals came to be known as less common metals or rare metals. In the process metallurgy of less common metals, the existing collection of processing techniques and the associated hardware that was so effective with the common metals proved inadequate. Generally, the less common metals cannot be processed while exposed to air. Either reduced pressure or a protected atmosphere is necessary. The temperatures needed are usually higher, and crucibles and other hardware should be chosen keeping their pronounced reactivity in mind. Rare metals, in order to be useful for their special functions, have to be in a state of very high material purity. Due to one or more of these factors, new techniques of processing had to be developed in metallurgy in the post-war era (Hampel 1961;Jamrack 1963; Winckler and Bakish 1971). Halide metallurgy, controlled atmosphere processing and vacuum metallurgy became the mainstream techniques. Considerable progress was achieved in all of these. For the rare metals, it was not just the ore processing and obtaining the metal that was different and difficult. Equally challenging has been the conversion of the as-produced metal into a usable form. That, incidentally, remains a work in progress to date.

The 20th century has been eventful in the evolution of materials and techniques, not only as regards the final products but also with respect to the technology enabling materials, a new name given to ceramics and refractories by Lee and Moore (1998). As crucible materials, the development of ceramics in the last 50 years has been remarkable, and their amenability to modification and enhancement has been the true reason behind the hope and success in many difficult situations of metals processing.

1.3   NON-FERROUS METALS

The word *smelting* is probably the most used term in any extractive metallurgy text. It represents the process by which a metal is obtained in the elemental form, or even as a simple compound, from its ore. Smelting involves heating beyond the melting point, usually in the presence of oxidizing agents, such as air, or reducing agents, such as coke. As noted earlier, copper was the first metal to be smelted in ancient times, followed by tin, lead and silver.

Over thousands of years of metalmaking, certain devices have evolved, in one form or another, as the key features of the smelting furnace. One is the tuyere, a pipe through which air can be forced into the smouldering coke, then the bellows or blower, an implement for forcing air into the tuyere, and the hearth, a place where the burning fuel can be contained over or against the tuyere opening. Traditionally, hearths have been built of mud-brick, fired brick, stone, or later, refractories of every description.

1.3.1 COPPER

The furnaces for producing non-ferrous metals and the small shaft furnaces used for ironmaking in the Roman period are similar. These were of square cross section, about 0.8 m on each side, and 2 m high externally. The main difference between the furnaces for different metals was in the hearths, in the arrangements for separating and tapping slag and metal (Tylecote 1992). A shaft furnace (Figure 1.5) is a general-purpose fixed bed reactor with an upright working chamber and is used to roast or smelt lumped materials. The heat required for the process is provided by the combustion of fuel, which usually is also a reactant. The charge introduced at the top of the furnace descends against the ascending stream of gases introduced in the bottom, resulting in direct contact between the charge and the hot gases. The molten products collect in the hearth at the bottom or flow out to a fore-hearth.

Pyritic (sulphide) copper ore was first roasted, in the open, over burning wood in shallow cavities in the ground. The roasting usually went on for over a month before completion. The oxide was then mixed with charcoal and fluxes and reduced in shaft furnaces. The ore was charged with alternate layers of charcoal into a furnace, and air was blown with bellows through the tuyeres. The furnaces were lined with a clay–charcoal mixture (refractory lining) rammed into place at the bottom and smeared on to the inner walls.

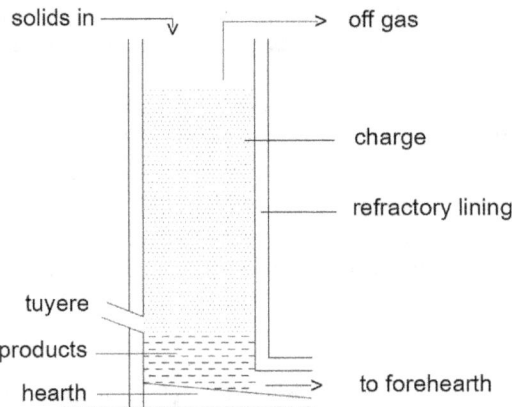

FIGURE 1.5 Shaft furnace.

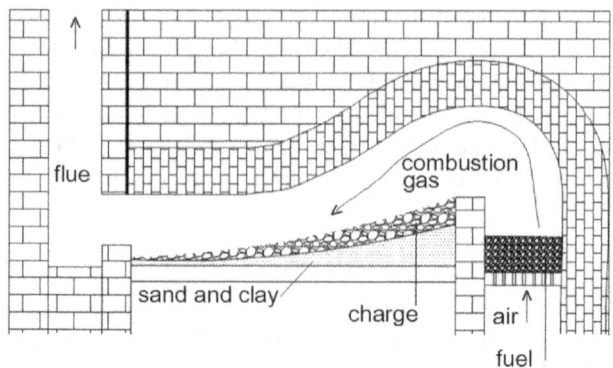

FIGURE 1.6   Reverberatory furnace.

The entire smelting process was rather elaborate, with repeated furnace treatments for days on end. Iron oxide, which was inevitably in the charge, was also reduced. Later, in 18th-century Europe, copper extraction from sulphide ores continued along the lines described (Tylecote 1992) as well as by the use of the reverberatory furnace.

Chronologically, the reverberatory furnace was developed later than the shaft furnace and the retort. In the shaft furnace, the fuel is blended with the charge to provide not only the heat but also reagents. In retorts, the charge is physically isolated from the source of heat and the products of combustion. The reverberatory furnace, represented schematically in Figure 1.6, comes midway between the two. A reverberatory furnace isolates the material being processed from contact with the fuel but not from contact with combustion gases. The term *reverberation* is used here in the sense of rebounding or reflecting of the flame and heat.

Reverberatory furnaces were first used in the medieval period to melt bronze for casting bells. They were used for smelting only in the late 17th century. Starting in about 1678, Sir Clement Clerke and his son Talbot built reverberatory furnaces for smelting lead and a decade later, for smelting copper and also tin. These units were 1.7–1.8 m long by 1 m wide in the inside. The hearth was made of sand with some clay, and the ore was fluxed with some lime.

In the second half of the 19th century, the major development in copper extraction was the use of the Bessemer process for the rapid oxidation of copper mattes, which had been made by concentrating and separating the sulphides in the blast furnace or a reverberatory furnace. The electrolytic refining of a slab of crude copper from the converter, invented by Elkington in 1865, was first successfully put into operation in South Wales in 1869. The modern flowsheet for copper extraction was thus in place.

1.3.2   BRASS

The term *brass* was used to denote brass as well as bronze in olden days. Even Biblical references to brass are known to actually mean bronze. This confusion persisted until recent times. Brass, an alloy of copper and zinc, was unknown in the

Egyptian and Babylonian civilizations, as well as to the Greeks and the Romans before the first century BC, while it was already in use in Palestine.

Starting from the Roman period, a process called cementation was used to make brass. Solid copper metal, zinc oxide or carbonate, and charcoal were placed in a crucible, which was heated to about 1000°C. Elemental zinc, which forms as a vapour by the carbothermic reduction of zinc ore at 1000°C, diffuses into metallic copper and converts it to brass. The ceramic crucible was kept closed with a lid to prevent zinc vapour from escaping before it had a chance to combine with copper. The use of porous ceramic vessels was important for the vessel to lose gas through the walls and thus prevent the build-up of excessive pressure. Incidentally, the process for carrying out cementation for brass did not change markedly until the 19th century. The ceramic used was a clay.

The crucible has an interesting function in the ancient process known as cupellation. Native silver exists but is not commonly seen. It is usually found in nature combined with other metals, e.g. in the lead minerals galena (lead sulphide) or cerussite (lead carbonate). Silver is produced primarily by smelting and then cupellation of argentiferous lead ores. Smelting yields silver and lead combined. Lead melts at 327°C, lead oxide at 888°C and silver at 960°C. To separate the silver, the alloy is melted in an oxidizing environment at 960°C to 1000°C. The lead oxidizes to lead monoxide, known as litharge. The liquid lead oxide is removed or absorbed by capillary action into the hearth linings. The base of the hearth was made in the form of a saucepan and covered with an inert and porous material rich in calcium or magnesium, such as shells, lime or bone ash (Bayley and Eckstein 2006). A calcareous lining is essential because lead reacts with silica (clay) to form lead silicate. Lead silicate is viscous and impedes the absorption of litharge. Lead does not react with the calcareous materials. It is captured by the materials purely by the physical phenomena of absorption. Some of the litharge may evaporate, but the rest is absorbed by the porous earth lining to form 'litharge cakes'. The litharge could be reduced back to lead separately.

Cupellation was known and used since the early Bronze Age. However the vessels (cupels) to carry out this process were first made only in the 16th century (Rehren 2003). The cupel is a porous container, usually made of bone ash or ceramic, and resembles a small egg cup. By the Middle Ages and the Renaissance, cupellation was not only the common process for refining precious metals but also a method for assaying precious metals. The process had remained virtually unchanged since its inception (Tylecote 1992). The cupellation furnace bottoms are separately smelted to recover the litharge and any silver contained in them. Cupellation is still in use.

1.3.3 TIN

In England, the tin ores of Devon and Cornwall were smelted in shaft furnaces (Tylecote 1992). In the 16th century, the furnaces for tin smelting were about 2.5 to 3 m high and 60×30 cm internally (Figure 1.5). They were constructed of large sandstone blocks and were lined with clay. The downward inclined tuyere was located at the back opposite the taphole. A sloping sandstone slab was the hearth, and the taphole was always open. The metal and slag separated out in the large (0.3 m long) fore-hearth, and the slag overflew from this. A charcoal cover was maintained over the tin to protect it from oxidation. The tin metal, smelted with water power and local fuel (probably peat), was re-smelted (or refined) and recast into ingots. The metal was ladled out of the fore-hearth and refined by drossing in an iron 'pot', similar to the 19th-century lead refining pot. After this, it was ladled into granite moulds. Charcoal was the main fuel. Smelting in a shaft furnace with charcoal followed by a limited amount of drossing to reduce the iron content yielded very pure (99.9%) tin. The reverberatory was not introduced here until 1700 CE.

1.3.4 ZINC

Metallic zinc is not an easy metal to produce because (i) zinc oxide cannot be reduced by charcoal at temperatures below 1000°C and (ii) zinc boils at 923°C, and the metal is a vapour at the temperatures required for reduction. The reduction temperatures of around 1000°C can be easily obtained by means of a bellows blast, but special arrangements were necessary to capture zinc vapour and collect the zinc metal. Zinc must be condensed to liquid metal before it comes into contact with enough air and is oxidized. This was the reason for the delayed appearance of zinc metal in the historical narrative of early metallurgy. Besides, given the ease with which calamine brass could be produced, there was not much incentive to use metallic zinc for this purpose.

In India, at Zawar near Udaipur in Rajasthan, zinc metal was produced in large quantities, beginning in the 10th century. Large heaps of small retorts were found there. They consisted of pointed elliptical masses of vitrified clay 25 cm long and 15 cm in diameter. These were all closed at one end but open at the other, with the remains of 2.5 cm tubes inserted in them. The heaps at Zawar suggest the extraction of about 100,000 t of metallic zinc (Darling 1990). A reconstruction of the production facility indicates that as many as 36 of these retorts were placed in a square furnace. Their openings projected through a square perforated plate such that the zinc vapour condensed and ran down into saucer-like vessels below. The retorts were charged with zinc mineral and charcoal.

In 1807, the Abbé Dony established a works at Liege, Belgium that used horizontal retorts, shown schematically in Figure 1.7, in horizontal rows (Darling 1990).

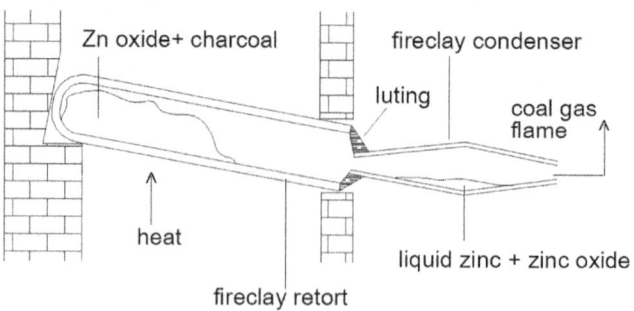

FIGURE 1.7   Horizontal clay retort for zinc production (From Darling, A.S. 1990).

A charge of ground calcined ore and coke was placed in the clay retorts. Zinc was collected in clay tubes fitted to the ends of the retorts after charging. As late as the mid-1950s, nearly half of the world's total zinc was produced in horizontal retorts of Dony's type. In the 1930s, the vertical retort furnace was developed by the New Jersey Zinc Company, and later, in the 1950s, the invention of the lead splash condenser made the blast furnace a practical and economic proposition for the recovery of zinc (and lead).

1.4 IRON AND STEEL

In the chronology of metalmaking, iron entered the scene a little later than copper and other major non-ferrous metals, but from then on, the metal and the techniques and materials associated with its processing had an overwhelming influence on metallurgy. A technique or material validated for the processing of iron and eventually, steel proved useful for other, non-ferrous metals as well with a few tweaks.

1.4.1 BEGINNING OF IRON SMELTING

When iron is smelted in the same way as copper or tin, the resulting metal is not in a form that is remarkably stronger or harder. As in the case of copper, easily smelted iron ores could have been the source of iron in the early days. The earliest iron produced was probably formed accidentally in fires burned while obtaining red ochre for use as paint (Bunch and Hellemans 2004).

Around 1500 BCE, the Hittites began smelting iron in Anatolia using clay-lined furnaces (Forrester 2016). Iron in malleable form was obtained by direct reduction of its ores with charcoal. Two types of furnace were used for iron making (Figure 1.8, Figure 1.9): bowl furnaces and shaft furnaces. Bowl furnaces were made by scooping out a hole in the ground and providing for air from bellows to be introduced through a tuyere. Stone-built shaft furnaces also used tuyeres but generally relied on natural draught. Originally, the process was carried out in pits lined with refractory

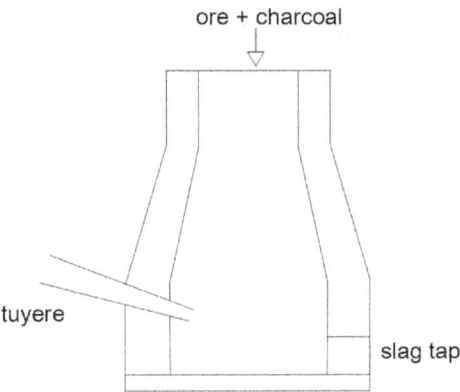

FIGURE 1.8 Bloomery.

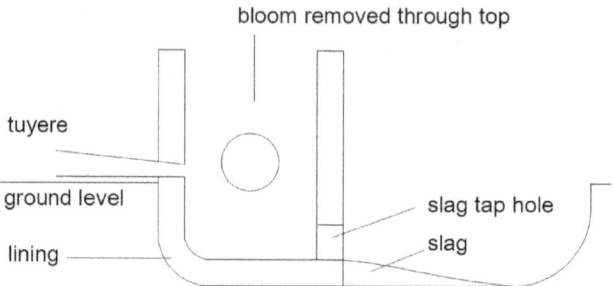

FIGURE 1.9   Roman bowl furnace (From Tylecote, R.F. 1992).

clay or on stone hearths. The bottom and sides of the hearth were lined with a thick coating of charcoal dust. In these furnaces, smelting was carried out by laying out a bed of red-hot charcoal on which iron ore mixed with more charcoal was added from above to obtain alternate layers of iron ore and charcoal. In such rather primitive arrangements, temperatures of only up to 1150°C could be reached. Reduction of the ore occurred, but the elemental iron formed was low in carbon and remained solid in the form of an incandescent bloom. Some slag and pieces of charcoal invariably remained entrapped in the bloom. The bloom needed to be pulled out from the hearth. By hammering this mass, the non-metallic constituents were squeezed out, and the metal was consolidated. The furnace for making this iron bloom is known as a bloomery. A bloom might have weighed about 8 kilograms during Roman times. This process is the oldest iron making process and has been in use, more or less uninterrupted, from as early as the second millennium BC until the early 20th century. The making of iron artefacts involved heating blooms in a fire and hammering the red-hot metal to produce tools and weapons. Iron obtained in this way is known as wrought iron.

Over time, the bloomeries were improved by using bricks and natural refractory stone for wall linings. A forced draught was supplied by bellows. Many bloomeries used low shaft furnaces, i.e. blast furnaces with a short shaft. Early furnaces were lined with natural sandstones or were even carved directly into stone ledges. By about 1300, higher temperatures were reached by using manually operated bellows to create the updraft in tall furnaces. By 1325, waterwheel-driven bellows had enabled a more powerful and sustained blast, which led to greater heat generation. With this improvement, bloom size could be increased. Blooms weighing over 100 kilograms were pulled out through the top of the shaft. The continuous operation enabled by the application of water power to bellows eventually made the blast furnace possible.

A popular version of the bloomery was the Catalan forge, an improved furnace that was invented around 700 CE in Catalonia and remained in use in Spain until the 19th century. The furnace, shown schematically in Figure 1.10, consisted of a small square cavity or hearth with an inclined copper tuyere designed to direct the blast onto the mixture of charcoal and ore. The nozzle of the device that provided the blast opened into the tuyere, thus drawing much more air on the principle of

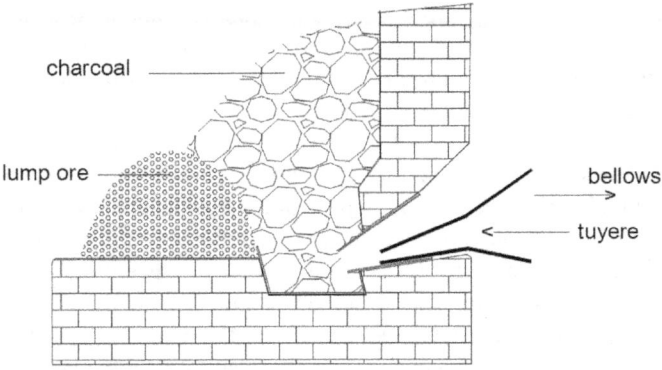

charcoal

lump ore

bellows

tuyere

FIGURE 1.10 Catalan furnace.

the injector. Another variant, the high bloomery furnace, had a taller shaft, e.g. the 3-metre high Stückofen. The blooms produced in this furnace were large and were usually removed through a front opening in the furnace.

1.4.2 BLAST FURNACE

The blast furnace is a vertical shaft furnace in which air introduced under pressure into the bottom of the furnace burns the fuel, usually coke, present in a mixture of metallic ore, coke and flux fed into the top. The ensuing smelting reactions produce liquid metal and molten slag, which collect in the hearth at the bottom and are recovered. Blast furnaces are used to produce pig iron from iron ore for subsequent processing into steel, and they are also used for production of lead, copper and other metals. Intense combustion is maintained by the current of air under pressure.

In Europe, the shaft furnace, the precursor to the iron blast furnace, developed gradually over the centuries from the small furnaces of Roman times. Increase in the height of the furnace, coupled with a continuous and strong flow of air sustained by mechanical bellows, resulted in the much higher furnace temperatures needed to produce the carbon-laden, lower-melting pig iron. The molten iron could be poured into moulds or could be remelted and cast into any shape. Thus, the shaft furnace of the bloomery metamorphosed into the blast furnace of the iron works. It was not until the 14th century that smelting furnaces capable of melting iron were built in Europe.

An incentive for the development of the blast furnace was the need for cast iron to make cannons. A cast iron gun was not only cheaper but also better than a bronze gun. Blast furnaces are necessary for producing cast iron. The first generation of furnaces (Tylecote 1992) were square externally, built of stone and lined with sandstone. This type remained prevalent up to the 19th century. The higher temperatures achieved by 17th-century blast furnaces led to the use of firebricks made from kaolinite-rich clays.

The main difference in operation between a bloomery and a blast furnace was in the tapping of liquid iron in addition to slag from the bottom of the furnace. A 2-m tall bloomery might be used to produce liquid iron by increasing the fuel/ore ratio to make the atmosphere more reducing and carburize the iron, which would then melt at about 1200°C instead of 1540°C. During this, the slag might become very viscous as its iron content was reduced, and it would not run easily. The addition of crushed limestone and maintaining a working temperature of at least 1300°C could result in a more fluid limy slag. Typically, pig iron has a carbon content in the range 3.8–4.7%, along with silica and other constituents of dross. As produced, it is very brittle and has little use directly as a material. Charcoal was the only furnace fuel until the 17th century, but its depletion and public concerns over the denuding of forests forced an eventual shift to coal. This changeover from charcoal to coke and later, the advent of the steam engine to power the air blowing cylinders of the blast furnace made much larger blast furnaces possible.

Throughout the 18th century, there was a steady increase in the size of the blast furnace. The maximum height increased from about 7 m in 1650 CE to 13.5 m in 1800 CE. By the middle of the 18th century, the shaft lining was usually brick, although the bosh and the hearth were made of refractory sandstone.. Brick was used for the stack in Maryport (1752 CE) and Low Mill (1761 CE). Moulded slag blocks were used to build the shaft of a Swedish furnace at Soderfors.

The early development of metallurgical blowing engines was one of the reasons why the Chinese gained early superiority in the cast iron tradition. The blast furnace was being used in China long before it became commonplace in Europe. The blast furnace was also in use in other countries of the East. The furnaces were made of sun-baked fireclay, sometimes from bricks and sometimes monolithic, with the better-grade material used near the hearth (zoning). The tuyere was of best fireclay. The unroasted ore was smelted with charcoal and without flux. The slag was free running and contained some iron.

Until World War II, the hearth walls were similar to a simple chimney-like structure built of fireclay brick. Many refractory combinations were tested. Until the 1960s, the hearth was composed of carbon wall and ceramic (fireclay) pads. This was further improved by using carbon hearth pads in addition to carbon hearth walls with a provision for cooling. In another arrangement, the combination was low-conductivity ceramic in contact with the metal but cooled by a carbon backing. The low-cost firebrick was eventually replaced with high-alumina brick.

The modern blast furnace is a tall, vertical shaft that consists of a steel shell with a refractory lining of firebrick and graphite. A schematic is shown in Figure 1.11. At the bottom is a parallel-sided hearth to collect liquid metal and slag. Above the hearth is the bosh, an inverted truncated cone. Air is blown into the furnace through water-cooled nozzles or tuyeres made of copper and located at the junction of the hearth with the bosh. The temperatures are the highest in the bosh and hearth. Here, the lining is usually made of carbon bricks that are manufactured by pressing and baking a mixture of coke, anthracite and pitch. Carbon is more resistant to the corrosive action of molten iron and slag than the aluminosilicate firebricks. For the remainder of the lining, aluminosilicate firebricks are used.

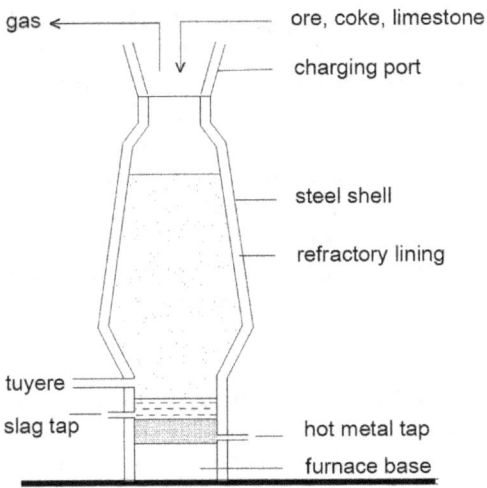

FIGURE 1.11   Iron blast furnace.

In the bosh parallel, bricks containing 63% alumina are used, and 45% alumina bricks are used for the stack.

During the 1960s and 1970s, the major changes in blast furnace linings were the replacement of fire clays with more refractory (higher-alumina) materials for the bosh and stack and a range of carbon materials for the hearth. Some of the new linings consisted of castables, silicon carbide and a thick ceramic layer for the upper portion of the hearth pad. The use of staves, which are 2 m × 1 m × 250 mm cast iron blocks embedded with steel pipes and ceramic inserts, also became common. In some of the large blast furnaces, the cooling and lining design relies on the combination of machined copper plate coolers and high-conductivity graphite.

1.4.3   Cast Iron

Cast iron is the name for iron–carbon alloys containing carbon in the range from 1.8 to 4 wt.% and silicon at 1–3 wt.%. Depending on its composition, cast iron melts at 1127–1204°C. Its properties include low melting temperature and good fluidity and castability. The oldest cast iron artefacts date to the fifth century BC, discovered in the present-day Jiangsu in China. Certain factors naturally favoured China being the earliest cast iron producer. The iron ore in China contained a high proportion of phosphorus, and residual phosphorus lowers the melting temperature of iron. They also had good supplies of refractory clay suitable to make kilns that could endure high temperatures. Furthermore, they invented and used a form of piston bellow, which provided a steady draught necessary for reaching and maintaining high kiln temperatures.

Cast iron was recognized as a superior material for making cannons. In Europe, cast iron production started only in the 13th century CE. The remelting of pig iron from the blast furnace in a reverberatory or air furnace was the process used to produce high-quality cast iron for making cannons in Britain in the mid-18th century (Tylecote 1992). Earlier, iron was cast direct from the blast furnace. The quality of the cast metal was determined by the smelting regime itself, which was difficult to control. When the raw pig iron from the blast furnaces was poor in quality and contaminated with slag, remelting in a reverberatory furnace enabled the slag to float out because a higher temperature was available. The melting conditions being slightly oxidizing, some of the carbon was also eliminated, giving a greyer and softer iron with high fluidity. Cast iron was also made by melting pig iron in cupolas. The cupola is another version of the shaft furnace, mainly used for remelting before casting.

1.4.4 STEEL

Even before the start of the Iron Age, people living in present-day Armenia chanced upon a technique to enhance the properties of iron. It was all meteoric iron at that time. They discovered that reheating and hammering iron produced a weapon material very much superior to copper or bronze. With this secret knowledge of making iron weapons, they enjoyed unequal advantage in their military campaigns across the Near East and the eastern Mediterranean (Bunch and Hellemans 2004). They had actually made weapons of steel, probably without realizing it, just as in later years, the makers of the famous Hessian crucibles used mullite as the material, probably without realizing it.

1.4.4.1 Cementation Steel

Steel is a mixture of iron and carbides of iron containing 0.5–1.5% carbon overall. One characteristic of steel, that the simple operation of heating it to redness and plunging it into water hardens it significantly, was a great reason to convert iron to steel. Cementation steel is made by a process that converts low-carbon iron to steel by letting carbon permeate into it. It may be recalled that in the context of brass, cementation refers to permeation of zinc into copper. Steel was sometimes produced in bloomeries when the temperature was high and charcoal was present in excess, conditions conducive to picking up of some carbon by iron. More often, it is made by heating iron in contact with carbonaceous mixtures. Even though exotic carbon sources like horn dust were used in the 10th century CE, the process had already been practised using more mundane materials many centuries earlier.

1.4.4.2 Crucible Steel Ancient Practice

Crucible steel was also made by melting together pig iron or cast iron, iron, sometimes steel, and often sand, glass, ashes and other fluxes in a crucible. In olden days using charcoal or coal fires, temperatures high enough to melt steel or low-carbon iron could not be reached, but temperatures high enough to melt pig iron were reached. The process therefore consisted of melting pig iron, and by soaking wrought iron or steel in the liquid pig iron long enough, the carbon content of the pig iron was reduced as it slowly transferred into the wrought iron. Crucible steel of this type was produced during the medieval era in south and central Asia. This steel was then forged, filed or polished to make weapons.

1.4.4.3 Puddling Furnace

In 1784, inventor Henry Cort unveiled his pudding furnace, which would convert pig iron to malleable iron. In a coal-fired reverberatory furnace, Cort melted a charge of pig iron, to which iron oxide was added as a slag former. The resultant 'puddle' of metal was agitated to speed up the removal of the impurities, carbon, silicon, phosphorus and manganese, by oxidation. The ensuing change of composition raised the melting point of the metal. While the slag remained fluid, the metal was rolled as a solid mass, formed into balls and taken out of the furnace. The slag was squeezed out of the ball by hammering. For a certain time period, puddling furnaces were able to provide enough iron to meet the demands for machinery and construction. Incidentally, the Eiffel Tower, which was constructed from 1887 to 1889, was made using puddled iron, approximately 7300 tons of it.

1.4.5 BESSEMER, OPEN HEARTH, LINZ–DONAWITZ (L-D) PROCESSES

Henry Bessemer announced in 1856a process for the production of steel from molten pig iron (Gale 1990). Another technique, the open hearth furnace for steelmaking, came into the picture a decade later, in1865. In the Bessemer process, the removal of impurities from the iron occurs by oxidation due to air blown through the molten iron. The oxidation, being exothermic, also raises the temperature of the melt and keeps it molten. An American inventor, William Kelly, had also asserted his independent discovery of the 'Bessemer process' back in 1851(Tylecote 1992)

Oxygen in the air blown into the pig iron melt oxidizes carbon and also the impurities silicon, manganese and phosphorus, to form the respective oxides. Oxides of carbon escape as gas and the other oxides are removed as the slag. The slag is composed of the oxidation products silica and manganese oxide, some iron oxide and also a certain amount of the refractory lining. The process was first carried out in the vessel shown schematically in Figure 1.12. The brick linings were made of pure

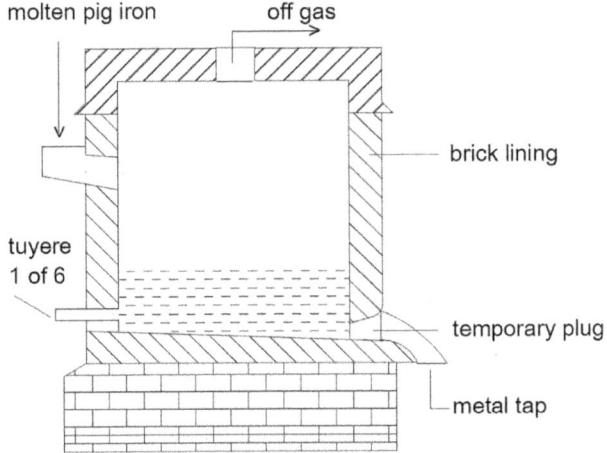

FIGURE 1.12   Bessemer's fixed vessel (From Tylecote, R.F. 1992).

sandstone or silica (SiO_2), the only materials then available for service at 1600°C for any length of time. Silica is chemically acidic and reacts on contact with the alkaline materials lime and magnesia. Bessemer, like all other furnace builders, had used siliceous materials for his linings, and these (acidic) linings would not combine with the acidic phosphorus pentoxide (P_2O_5) that forms during the oxidation of phosphorus. Thus, phosphorus was not removed from the steel. This situation confined the use of the Bessemer process to only non-phosphorus ores, i.e. those containing less than about 0.05% P. The ores that accounted for most of the iron production of 19th-century Britain had phosphorus content exceeding this limit and were unsuitable for the Bessemer process. For solving the problem of phosphorus in steel, neutralization of the acidic P_2O_5 with lime was considered. The basic slag so produced tended to consume the siliceous lining also. The need for a basic lining was strongly indicated.

In 1877–1888, Sydney Gilchrist Thomas and Percy Gilchrist checked the effectiveness of a basic lining using an 8-lb converter lined with limestone bonded with sodium silicate (Lee and Moore 1998). With this lining and a slag made basic by adding limestone, phosphorus was removed from the metal and poured off with the slag. They then developed a lining of fired dolomite mixed with tar, which was operated successfully in the larger converters at Middlesborough in 1879. The first Bessemer furnace was built in the United Kingdom in 1884; it had a tarred dolomite hearth. During operation, there is considerable movement of the entire vessel through the positions shown schematically in Figure 1.13.

1.4.5.1 Open Hearth Furnace

The open hearth furnace (OHF) was developed by Carl Wilhelm Siemens and Pierre-Émile Martin, and the process is also known as the Siemens–Martin process. The open hearth process is another method to oxidize carbon and other impurities out of pig iron to produce steel. The OHF itself consists of a shallow, rectangular hearth, like the hearth of a reverberatory, to hold the charge, the liquid steel and the slag,

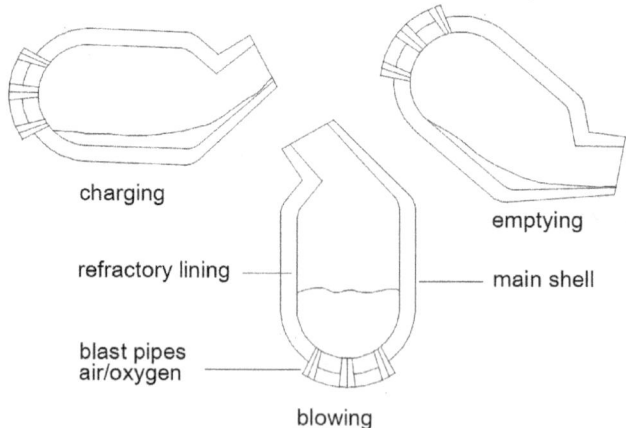

charging

emptying

refractory lining — | — main shell

blast pipes
air/oxygen —

blowing

FIGURE 1.13   Bessemer converter positions.

relying on external heating that uses the heat of combustion of gaseous or liquid fuels by preheated combustion air. Typically, the OHF is first charged with light scrap, and heated using burning gas. Once the charge has melted, heavy scrap is added, together with pig iron from blast furnaces. Once all the charge has melted, limestone and other slag-forming agents are added. Iron oxide and other impurities provide the oxygen to burn away the carbon in the pig iron. To provide more oxygen in the heat, iron ore can also be added.

This is a batch process, much slower compared with the Bessemer process and thus, easier to control. It can be terminated when the target carbon content has been achieved. Attempts were made to address the phosphorus problem using basic refractories. The basic OHP was first operated at Le Creusot, France, in 1880. However, the OHP was not very effective at removing all the phosphorus.

Apart from the phosphorus problem, the brittleness caused by nitrogen in the blowing air prevented Bessemer steel from being used for structural applications. Nitrogen was not an issue in the steel produced by the OHP. Open hearth steel was suitable for structural applications, and the process started gaining acceptance in the year 1865. For about 100 years, the open hearth and Bessemer-based processes jointly contributed to most of the steel that was made. Then came the L-D and electric arc furnaces, which represent the current steel making technologies.

1.4.5.2 Linz–Donawitz (L-D) Steel

Doing away with the bottom tuyeres, the L-D process uses a vertically positioned lance, lowered from above, which is able to blow oxygen onto the surface of the molten metal. This process was developed in Austria by the Linz–Donawitz companies following the initial demonstration of its working by Robert Durrer and Heinrich Heilbrugge in 1948.

Basic oxygen steelmaking is more similar to the Bessemer process (decarburization by blowing oxygen as gas into the melt) than to the open hearth (burning the excess carbon away by adding oxygen-bearing substances into the charge). The L-D process simultaneously addressed both the phosphorus and the nitrogen issues by having a basic refractory lining and using oxygen blow instead of air blow. The L-D process is often referred to by other names. Basic oxygen steel making (BOS), basic oxygen process (BOP), basic oxygen furnace (BOF), oxygen steel making (OSM), oxygen converter process are other names for the Linz–Donawitz (LD) steel making process.

1.5 REFRACTORIES IN THE 20TH CENTURY

Certain naturally occurring inorganic materials have been used as crucibles or in lieu of crucibles for metals processing since the Bronze Age. Locally available clays, stones, river sand, sandstones, granite, fire clays and fire bricks have all been used at various stages, and these were all that were generally accessible for use up to the post-medieval period. This rather ordinary-looking inventory of crucible hardware was at times punctuated by the appearance of special refractories: a refractory consisting of steatite ($Mg_3(Si_4O_{10})(OH)_2$) in the Bronze Age, the Hessian crucibles made of mullite and even graphite in the post-medieval period. Systematic and directed development of materials for crucible functions coincided loosely with the industrial revolution, when furnaces that started operating at consistently higher temperatures became common. The rapid developments of the 20th century in the production of metals and many other types of materials have been closely interlinked with concurrent developments and advances in the refractories.

1.5.1 SILICATES

Up to the end of the 19th century, there was a heavy reliance on natural materials, such as rocks and fireclays, for use as refractories. Because of their variable composition and properties, the predictability of their somewhat modest performance was also limited. Fireclay was probably the first fired refractory to be used. Being plastic when mixed with water, it could be readily shaped into crucibles convenient for small batch melting of metals at moderate temperatures. Clay-derived aluminosilicate refractories are acidic in their chemical nature and therefore, were useful to contain acidic slags and fluxes reasonably well. They proved mostly adequate during the early industrial revolution. Silicate was the first lining material for the Bessemer converter. In England, crushed Dinas rock from Wales was utilized from 1822 to form nearly pure silica brick cemented together by the smallest possible amount of lime as binding flux. These bricks, developed by William Weston Young, were the most suitable refractory available for lining acid open hearth furnaces, where temperatures were much higher than in Bessemer converters (Lee and Moore 1998). The temperature capability of the refractories was stretched to the higher side by increasing their alumina content. The reason was highlighted in 1924 when the binary Al_2O_3–SiO_2 phase diagram was determined by Bowen and Greig (1924).

The 1920s and 1930s can be regarded as the golden era of refractory science. The development of X-ray diffraction (XRD) techniques for elucidating the crystal structures of clays and oxide–silicate systems, and the determination of the important binary and ternary phase diagrams, happened during this era. The publication by Hall and Insley (1933) of a compilation of phase diagrams, apart from being an invaluable resource, is also symbolic of this renaissance.

1.5.2 DOLOMITE

The first use of dolomite as a refractory was in 1878, when Sydney Gilchrist Thomas lined his experimental Bessemer converter with tar-bonded, burned dolomite. The production of burned (fired) dolomite for the steel industry began in 1901. Dolomite was used in the form of metal-cased magnesite for open hearth and electric arc furnace (EAF) walls and as unfired, chemically bonded magnesite bricks until about 1930.

1.5.3 CHROME–MAGNESITE

It was becoming known in the 1930s that bricks made from graded mixtures of chrome ore (chromite, $(Mg,Fe)(Cr,Al)_2O_4$) and dead-burned magnesite showed synergistic improvement in properties. For instance, refractoriness under load of chrome–magnesite was much higher than that of either material separately. Silica roofs in all basic open hearth furnaces were eventually replaced by the superior chrome–magnesite roofs.

Chrome refractories are chemically neutral. They were used in basic open hearth furnaces later in the 19th century. In direct-bonded chrome refractories, the silicate films, which surround the chrome ore particles and periclase grains in the conventional brick, are replaced by direct periclase–spinel and periclase–periclase bonds. In this form, they have much improved hot strength, more resistance to slag attack and greater dimensional stability.

With the increase in the number of BOF installations, standard basic bricks containing magnesia, dolomite and chromite aggregates and matrix powders containing tars and pitches were developed. These were referred to as 'tar-bonded' and 'pitch-bonded' basic brick. In early 1950s, pitch-bonded dolomite and later, magnesia-enriched tar-bonded dolomite were developed for the BOFs. The original purpose of tar in refractories was as a binder and a coating to protect the basic grain, which is prone to hydration. It was soon realized that under the conditions of BOF steelmaking, magnesia and dolomite linings lasted longer if they contained carbon. Unfired pitch- and tar-bonded magnesia containing additions of graphite came into universal use.

The use of magnesia refractories in combination with carbon started in the early 1970s (Biswas and Sarkar 2020). A decade later, the development of resin-bonded magnesia – graphite, first with higher carbon content and then with the addition of antioxidant, further improved the life of BOF linings. Late in the 1970s, a product containing phenolic resin as a cold bond for a mixture of periclase, graphite and metallic antioxidants came to be used in Japan. Resin-bonded basic brick not only has superior properties vis-à-vis the tar-bonded variety but also is more acceptable environmentally. Fused magnesia with larger size crystals than sintered magnesia was recently introduced to further enhance the corrosion resistance.

The oxide–carbon–metal composite bricks are used in EAFs and BOFs and also in ladle slag lines. The oxide confers oxidation resistance, and graphite provides increased thermal conductivity, non-wettability and slag resistance. The carbon bond is developed in situ in these refractories on firing them in the furnace. The resin or pitch decomposes, yielding a fine pyrolytic carbon bond.

1.5.4 GRAPHITE

Natural graphite was first used as a refractory material in crucibles in the 15th century in Bavaria. Gousse Bonnin obtained a patent in 1769 for a crucible made of equal parts of clay and graphite (termed black lead or plumbago). The manufacture of plumbago crucibles started in the United States in 1827 and in the United Kingdom in 1856. Graphite use became more widespread in the 19th century with the discovery of more deposits.

Edward G. Acheson synthesized graphite accidentally while he was experimenting on carborundum, and he started commercial production of graphite in 1897. Since 1918, graphite, 99 to 99.5%pure, has been produced using petroleum coke, small and imperfect graphite crystals, and organic compounds as the raw materials. The use of graphite has since only increased both in variety and in quantity.

1.5.5 GRAIN AND BOND PHASES

During the past 100 years, in the usage of refractories, the trend in the selection of raw materials has shifted from the natural sources used in the nearly as-mined condition (clays and other aluminosilicates) toward more synthesized, higher-purity materials (fused and tabular grain and complex bond phases).

Up to the 1940s, grain phases used in refractories consisted of crushed firebrick, quartzite, pyrophyllite, calcined clays (grog), sintered magnesite, chrome ore, sillimanite–kyanite–mullite and calcined bauxite. From the 1940s, tabular alumina and fused grains of various oxides also came into use.

The production of refractories by fusion in an EAF or electrocasting began in 1902, when Carborundum Company produced alumina via this route. The Norton Company produced a ceramic-bonded fused alumina crucible in 1906. Refractories like magnesia and mullite are also made by fusion. They are also used as grain material in the crushed form.

In 1934, tabular alumina was made in a shaft furnace by T. S. Curtis. The process involved heating of 2-cm diameter pellets of calcined alumina at >1925°C, just below the melting temperature (m.p. of alumina is 2072°C) until almost complete conversion of the fine α-Al_2O_3 crystallites to large (50 to >900 mm), angular, elongated, tablet-shaped crystals. Tabular alumina crystals are hard and dense, and have good thermal conductivity and high crushing strength. The crystals contain substantial closed porosity because of the large grain growth, leaving pores in the centre of grains where volume diffusion is slow. This is one of the reasons for the better thermal shock resistance of tabular alumina than fused alumina. The closed pores act as crack arresters. Due to properties like dimensional stability, corrosion and erosion resistance, and load-bearing capacity, the use of tabular alumina in iron, steel, aluminium and other non-ferrous metal production has seen a steady rise since the 1970s.

1.5.6 BRICKS AND MONOLITHICS

Up to the 19th century, refractory brick was manufactured manually, using mainly aluminosilicate materials. Between 1900 and 1920, machine pressing came into use. Over the years, brick production was further automated, resulting in bricks more uniform in shape, having a more isomorphous microstructure and better properties. In the 1970s, carbon-bonded bricks were developed. These are fired in situ. Installation procedures were developed for monolithics for linings in the latter half of the last century.

1.5.7 REFRACTORY BONDS AND BINDERS

The names of refractories often indicate the type of bond that holds the aggregate/ grain phases together, e.g. carbon bonded, silicate bonded, phosphate bonded, cement bonded and so on. For a long time, the fired integrity of refractories relied on silicate or aluminosilicate bonds derived from vitrification reactions. During the 20th century, bonds were synthesized first through the use of fine additions, called mineralizers, and later on by the introduction of separate cementitious and glassy bond systems.

The significant development associated with manufacturing technology for linings in the second half of the 20th century has been the procedures for installation of monolithic or unshaped refractories that can be formed and shaped in situ. Monolithic refractory materials consist of mixtures of aggregates and binders, ready for use either as supplied or after the addition of suitable liquids. Since the 1960s, their use has become widespread as castables, gunning mixes, and mouldables or plastics. Unshaped refractories use similar aggregate materials as shaped brick, but the binders used to form the bonding phase are very different. The binders may consist of aluminate, silicate and phosphate bond systems or their combinations. The development of techniques for installation continued into the 1980s.

1.6 ELECTROLYTIC PROCESSES

At the very beginning of the nineteenth century, Sir Humphry Davy had stated that even the most stable chemical compounds should be electrolytically reducible with the aid of the newly available voltaic cell, and supported his assertion by obtaining sodium, potassium, barium, strontium and calcium metals using this approach. However, he did not succeed in his attempts to obtain the element he had only named 'aluminium'. An impediment was the unavailability of 'heavy' electric current.

1.6.1 ALUMINIUM

Generators capable of producing heavy electrical currents become generally available by the mid-1870s, and interest in electro-metallurgical possibilities revived. The initial attempts in the United States to obtain aluminium by the electrolysis of fused salts were made by Charles S. Bradley of Yonkers, NY. Even though his ideas and approaches were remarkably similar to those of Hall and Héroult, Bradley's patent was granted only in 1892, by which time the Hall–Héroult process was already in commercial operation in several countries (Darling 1990).

In 1885, Charles Martin Hall began to investigate various electrolytic approaches to aluminium production in his private laboratory and concluded that fused salt baths would be essential. On 10 February 1886, he also found that alumina readily dissolves in fused cryolite 'like sugar in water' and that the alumina/cryolite solution thus obtained is a good electrical conductor. It was, however, already well known that cryolite would dissolve alumina. By dissolving 15–20% of alumina in cryolite, Hall obtained a bath whose melting point was between 900 and 1000°C. At this temperature, its electrical conductivity was high enough to permit electrolysis. The cell used by Hall is shown schematically in Figure 1.14 (Darling 1990). Hall found that the bath rapidly dissolved silica also from the refractory materials used to contain

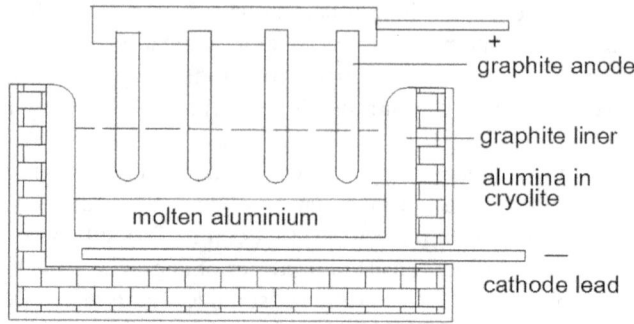

FIGURE 1.14 Hall aluminium cell (From Darling, A.S. 1990).

it. He soon solved this problem by containing his melt in a graphite crucible and obtained a number of small globules of aluminium, which formed close to the crucible that also acted as the cathode.

On 23 April 1886, Paul Louis Toussaint Héroult applied for a French patent for an invention which was identical to that of Hall. He described:

> a method for the production of aluminium which consists of the electrolysis of alumina dissolved in molten cryolite, into which the current is introduced through suitable electrodes. The cryolite is not consumed, and to maintain a continuous deposition of metal it is only necessary to replace the alumina consumed in the electrolysis.

Héroult did not realize initially that the bath required no separate external heating, and corrected the lapse in a subsequent patent. His original cell arrangement, shown in Figure 1.15 (Darling 1990), does show furnace heating of the cell.

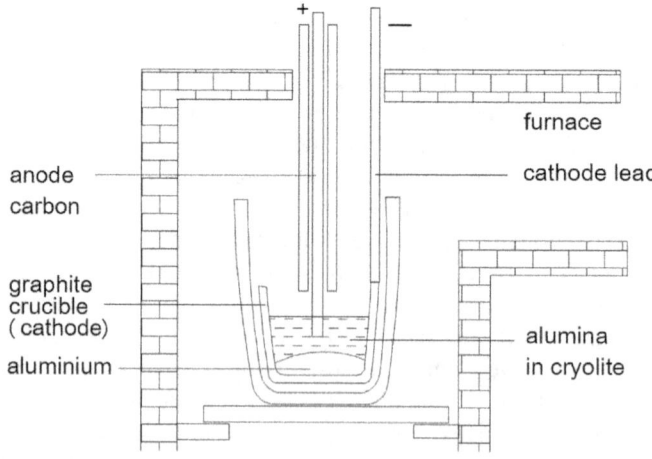

FIGURE 1.15 Héroult aluminium cell (From Darling, A.S. 1990).

1.7 RARE METALS

World War II affected major branches of technology in a profound but positive way, accelerating progress in general and causing the initiation and intensification of activities that would have, in normal times, remained peripheral. Developments were powered by atomic energy and the aerospace and electronics industries. All three of these were materials driven, and most of those materials were hitherto undeveloped or underdeveloped. Clifford Hampel, the author of the 1961 classic *Rare Metals Handbook* (Hampel 1961), covers as many as 54 metals in his handbook, implying their rarity before World War II and newfound relevance in technology in the postwar period. That was the status and background at the beginning of the second half of the 20th century. He mentions the story of uranium, noting that as late as in 1940, only a few grams of good uranium were in existence and that the data on the metal was so scanty that its melting point was listed only as $<1850°C$. The real value, as is known now, is $1132°C$.

Apart from the fact that they had been already discovered and also investigated as scientific curiosities, all relevant details of their production in required purities and sufficient quantities remained to be either developed or validated in the post-1940 period. These activities gathered momentum, and much was accomplished in the first three decades of the second half of the 20th century.

1.7.1 Special Crucibles and Protected Environment

The rare metals are produced by techniques that are in many ways different from those used for common non-ferrous metals and also iron and steel. More often than not, the processing of rare metals has to be carried out in a protected atmosphere and in inert crucibles. This is due to their pronounced reactivity with elements like oxygen, nitrogen, hydrogen, carbon and silicon, especially at the high temperatures needed for processing. Many of the rare metals are excellent getters and avidly pick up these impurities from the surrounding atmosphere. The metals also ingest them from any solid, such as the crucible, with which they are in contact in molten condition during processing. It so happens that very often, this aspect of rare metal processing is the decider that qualifies a process for production or processing. The metal–crucible interactions play a crucial role in the metallurgy of many rare metals.

Unlike the common metals, where the as-mined ore, after mineral processing operations for separating it from gangue, concentration and roasting, is taken for smelting, the rare metal minerals are put through very elaborate mineral processing and separation operations, and a pure compound is taken for reduction. This is because slag has literally no role in metal purification here, unlike with the common metals. Impurities that enter the as-reduced rare metal, barring a few well-known exceptions, cannot be removed from the metal sufficiently or easily. A certain similarity can be perceived with aluminium extraction. Bauxite is put through the Bayer process, and pure alumina is taken up for electrolytic reduction. Most of the metals presently produced by electrolysis appear to be designed to be made that way only.

Any attempts to produce them by alternate means have considerable technological and economic penalties.

Electrolysis in a fused salt is a potentially useful route for the preparation of metals that are not easy to prepare by a chemical reduction route. There are many examples. Aluminium, magnesium, titanium, zirconium, and the rare earth metals are some of the important cases. The success of molten salt electrolysis as a commercial route to produce aluminium and magnesium encouraged the intensive search for a similar process for zirconium and titanium. The challenge lies in the melting points of the metals produced. Al melts at 660.5°C and magnesium at 650°C. Fused salt electrolysis of these, carried out at 950–960°C for aluminium and at 700°C for magnesium, yields metal in a liquid form, which is tapped or otherwise collected conveniently without any problem of contamination. The melting points of zirconium and titanium are 1855°C and 1670°C, respectively. At least at the time when electrolysis was attempted for them, there was no intention of carrying out electrolysis at temperatures exceeding these values. The metal at the cathode will have to be in the form of a solid.

Even though the temperatures needed for most of the electrolytic reduction processes are lower than about 1100°C, there are often serious contamination issues amongst the metal produced, electrolyte and cell materials.

As well as electrolytic processes, metal replacement reactions or metallothermic reductions have been widely useful for the preparation of the rare metals. Even though some the processes were first demonstrated in the previous century, much of the usable information on process details was generated in the 20th-century post-war period.

1.8   METAL REPLACEMENT (METALLOTHERMIC) REACTIONS

Metal replacement reactions, also known as metallothermic reductions, are carried out in many distinct ways, and there are important examples for each of them.

1.8.1   Goldschmidt Process or Bomb Reduction

Towards the end of the 19th century, Johannes Wilhelm Goldschmidt invented a process whereby aluminium metal is oxidized by an oxide of another metal, which is in turn converted to the metallic form. The oxide originally experimented with was iron oxide, and the reaction occurs with an intense evolution of heat (highly exothermic). It is known that Goldschmidt was originally interested in producing very pure metals by avoiding the use of carbon in smelting. Events that led to the use of his process in the 20th century indicate that he eminently succeeded in his objective. The reaction is called the thermit reaction, and the process is called the Goldschmidt process (Goldschmidt and Vautin 1898).

The Goldschmidt process raises the temperature of the reaction mixture to high levels very quickly, and the reaction mixture would need to remain molten for some time in order for the molten metal to separate cleanly from molten slag. These two characteristics immediately point to an important role for any crucible in which the process is carried out. It is performed in a refractory-lined reactor of a suitable configuration. In order to prevent pick-up of interstitial impurities like oxygen and nitrogen from the surrounding air by the metal when it is at a high temperature and in molten form during the process, the thermit process is carried out under an inert gas cover. A refractory-lined and closed reactor is therefore the standard equipment for the Goldschmidt process.

As an example of direct application, pure, carbon-free niobium metal can be produced by aluminothermic reduction (Wilhelm et al. 1966) of niobium pentoxide. The same applies to vanadium metal also (Carlson et al. 1966). Not only vanadium metal but vanadium–aluminium master alloys have been produced routinely for use in the manufacture of Ti6Al4V alloy, a commercial alloy needed in high volumes. These processes were actively developed during the 1960s and 1970s. Wah Chang corporation at Albany used the process industrially beginning in the 1970s (Wang et al. 1970). A schematic of the closed bomb reduction reactor for commercial vanadium thermit (V-Al-O alloy) productionis given in Figure 1.16. Perfect (1967, 1981) of Reading Alloys introduced the revolutionary concept of using a water-cooled copper reactor, instead of a refractory-lined reactor, for conducting the thermit reaction. Carlson et al. (1981) of Ames Laboratory showed that using water-cooled reactors for aluminothermic reduction of vanadium can markedly improve the purity of metal that is produced. Examples of processes that are variants of the Goldschmidt process include the techniques for the production of uranium and plutonium metals. The differences are that a fluoride, instead of an oxide, is used as the start material, and the reducing agent is not aluminium but calcium or magnesium.

1.9 CONSOLIDATION AND REFINING

In the as-reduced condition, the rare metals are generally unusable. They need to be consolidated and also usually refined before conversion to mill products for use in fabrication. While the water-cooled copper reactor was first used in the metal reduction stage as recently as in the 1980s, such arrangements were in use in metal

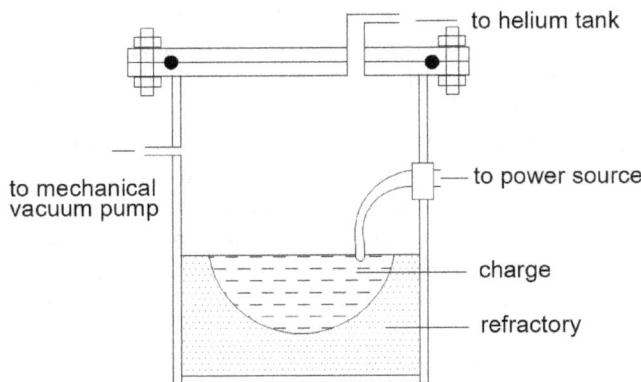

FIGURE 1.16 Closed bomb reactor for aluminothermic reduction (From Wang et al. 1970).

refining systems much earlier. Water-cooled copper moulds for holding the melt are key components in vacuum arc and electron beam furnaces. It has also been incorporated in induction melting, particularly in the induction skull melting configuration (Winckler and Bakish 1971)

Metals like tungsten and molybdenum are best consolidated by powder metallurgy techniques of compaction at high pressures and sintering at high temperatures. Melting is not very attractive, not only because of their very high melting points and absence of a suitable crucible if vacuum induction melting (VIM) is selected, but also because these metals do not cast well. Most of the other metals can be melted and generally cast well.

Melting to consolidate (with little or no refining) is always done in arc furnaces. Some of the metals and alloys are best melted (for refining and consolidation) in an Electron Beam Furnace (EBF) if there are no volatile components to be retained in the melt stock, because in electron beam melting (EBM), refining by vaporization also occurs along with consolidation. Except that the mechanism of heating is same, the construction and function of arc furnaces for rare metals processing is very different from that used, for example, in steel making.

In all processing operations on rare metals, the common and significant feature is the use of a protected atmosphere, either clean vacuum or an equally pure inert atmosphere. Processing equipment is constructed accordingly.

1.9.1 Vacuum Induction Melting

Vacuum melting, or melting a substance while it is in vacuum, is a 20th-century phenomenon (Winckler and Bakish 1971). The method of heating used here is only electrical and confined to the use of resistance, induction, arc and electron beam. Vacuum melting made the transition from the laboratory to the plant in 1917 when Dr. Wilhelm Rohn melted nickel base alloys by resistance heating. Although resistance-heated melting furnaces were used initially, the technique has never been preferred for production scale melting. This is due to several factors. Supplying heat from an external source is thermally inefficient. The thermally insulated crucible cannot be used, and the crucible is made to withstand temperatures in excess of that of the bath throughout its wall thickness. Outer wall temperatures several hundred degrees more than that of the bath are incurred when effective (practically useful) heating rates are used with refractory oxide crucibles.

E.F. Northrup built the first prototype of a vacuum induction furnace in the United States in 1920. Basically, in an arrangement for induction heating, a coil surrounds a piece of metal. When an alternating electric current passes through the coil, eddy currents are induced in the metal (charge). The resistance of the metal to the eddy currents raises its temperature. In place of a solid metal it is often a metallic melt suitably contained. By 1923, vacuum induction furnaces were commercially available from Heraus in Germany. They were used to produce thermocouple materials and resistance heating alloys.

In the VIM process, virgin alloys or scrap are melted down under vacuum or inert atmosphere to avoid oxidation and contamination from atmosphere. The molten charge is then cast, mostly under vacuum. With small furnaces up to 25 kg capacity, it is standard practice to use pre-fired crucibles. For larger sizes, the economics of rammed linings becomes more clear. An essential feature of VIM is the presence of a crucible, which may never be completely inert to molten charge.

Commercial VIM developed rapidly with the advent of titanium in the United States in the early 1950s and by the need for VIM of wrought nickel base alloys for turbine blades. The next good thing in the development was the availability of much improved vacuum pumps originating from the atomic energy industry.

1.9.2 Vacuum Arc Melting

The arc furnace is the first metallurgical furnace to work on electricity. The first arc furnace was developed by Sir William Siemens in 1878–1879. In vacuum arc remelting (VAR), heat is generated by the application of a low-voltage high-current electric arc to the surface of the molten metal from an electrode. Vacuum is in the range of 0.1–10 Pa. The electrode may be consumable or non-consumable. In order to assist ionization and also to minimize vaporization of the electrode, melting is usually carried out under a low pressure of inert gas and not in high vacuum. In the consumable electrode mode, as arcing proceeds, the electrode is consumed as it heats up, melts and drips into the metal pool (Winckler and Bakish 1971). Titanium sponge is consolidated to titanium ingot in this way. In the non-consumable mode, as the name implies, the electrode remains intact, and metals fed into the arc melt and collect into a water-cooled copper crucible.

In the standard consumable electrode arc furnace, the normal deep round crucible used for ingot production may be replaced with a shallower water-cooled crucible capable of being tilted through 90° so that its contents may be poured into conventional moulds to make shaped castings (Clites and Calvert 1961).

The term *skull melting* implies the use of a water-cooled copper crucible and denotes the fact that a thin shell, or skull, of metal remains behind in the crucible after pouring due to the rapid chilling effect of the water-cooled crucible. After its formation, the skull remains in the crucible, separating the melt and the crucible inner surface. Depending upon the thermal constants of the metal being melted and the power available, up to 90% of the crucible content can usually be poured. The principle has obvious applications in the manufacture of castings in the reactive metals, which do not lend themselves to melting in ceramic crucibles. In the field of titanium and titanium alloy castings, the method has been most widely adopted. However, little superheat of the metal is possible using a consumable electrode technique, since generally speaking, higher melting power merely results in quicker melting rates, and metal temperatures in excess of 30–40°C above the melting point are difficult to achieve.

Compared with the conventional DC machines, the AC system offers a better heat distribution and a shallower pool for a given melting current, since no current passes through the crucible, and the danger of penetration is minimized. Large homogeneous ingots, 700 mm in diameter, of certain superalloys prone to segregation have been made by Cooper et al. (1965).

1.9.3   ELECTRON BEAM MELTING

Electron beam melting (EBM) utilizes the kinetic energy of highly accelerated electrons, which are made to impact the material to be heated or melted. Heating or melting is a consequence of transfer of kinetic energy of the electrons to the lattice of the material on impact, leading to local temperature rise. A greater number of faster electrons will lead to higher temperatures.

The concept was first propounded during or about 1905, and its inventor was M. von Pirani. However, the technique became a useful tool only half a century later (Winckler and Bakish 1971).

Electron beam has truly outstanding advantages as a heat source for melting. It is clean, does not have any real upper temperature limit and works with high vacuum as its natural environment. The last advantage is sometimes cited as its limitation also. In a different format, electron bombardment heating, where the beam is not really focused but only directed, also has many applications as a heat source and is amenable to excellent and close temperature control.

Electron beam melting and purification owes its success to the ability to fuse any material known to undergo solid/liquid transition and to maintain the same in a fused state for as long as is necessary to effect the purification required. The melting is carried out in a water-cooled copper hearth, the popular and reliable non-contaminating crucible.

1.10   CRYSTAL GROWING

A melt, on normal cooling, solidifies to a polycrystalline solid. The nature and properties of the solid are in many ways determined or influenced by the details of these crystals and their arrangement. The technique of 'crystal growing' deals with the possibility of changing these details and arrangements on purpose.

Two major methods of crystal growing exist: techniques where the melt is contained in a crucible and techniques in which the material is self-contained, i.e. by surface tension.

1.10.1   CRYSTAL GROWTH IN A CONTAINER

The technique was actively investigated in the 1960s. Apart from the obvious role of the container in the shaping of the crystal, a number of physical and chemical properties of the container material are important. The container must be inert to the metal or alloy being purified. Wetting of the container by the melt should be minimized, because good wetting leads to sticking and eventual fracture of the container. The coefficient of thermal expansion of the container and the charge should be carefully matched. In order to minimize heat transfer problems, the thermal conductance of the container material should be comparable to or less than that of the charge. These general criteria allow a wide range of possible container materials. The choice becomes more limited as the temperature and/or reactivity of the molten charge increases.

1.10.2 Floating Zone Technique

Chronologically, investigations of the floating zone technique started simultaneously with the studies of the regular crucible techniques, in the 1950s and 1960s. In fact, there was more research activity on the floating zone techniques in the 1950s than on the other methods. The novelty value and the tremendous popularity of the reactive and refractory metals probably drove these early investigators. In float zone crystal growing, a molten zone of the material is held in place by its own surface tension between two collinear solid rods of the same material. The technique was originally developed by Keck and Golay (1953), Emeis (1954) and Theuerer (1956)to prepare high-purity single crystals of silicon. This method avoids crucible contamination. The material usually solidifies behind the molten zone to form a single crystal along the entire length of the specimen rod. Common heating methods for this are induction and electron bombardment.

Induction heating makes it possible to work under high vacuum, in a reducing or oxidizing atmosphere, or in a positive pressure of inert gas. Induction is particularly attractive for the growth of crystals of the reactive metals. Electron beam heating is possible only in a vacuum. It was first used successfully by Calverley and Lever (1957). Control of the dimensions of the molten zone is easier than with induction heating because the electrons can be focused on the specimen.

1.11 SUMMARY AND SCOPE OF CRUCIBLE–METAL INTERACTIONS

In a metallurgical process with materials placed inside a crucible, when heat is supplied to raise the temperature (or withdrawn to lower it), and substances necessary for chemical reactions are introduced, the objective is to induce changes to occur. Changes are of two types: those that are designed to occur, and those that are not intended and usually not wanted either. In general, all of these occur, governed by thermodynamic properties of the system and also, in reality, determined by the kinetics of the processes. Among the most important of these reactions and processes are those that involve the crucible material itself.

The general expectation is that the crucible material will remain inert through all these changes. Apart from physically containing the reactants and products and allowing or not allowing heat to pass through it, depending on the original purpose of its choice, it should remain indifferent to its contents and also to processes occurring within its confines. This is the perfect crucible. But, perfect crucibles do not exist.

All crucibles interact with their contents. They interact chemically and also physically. Interactions cannot be avoided but can be managed. Interactions can be generally understood, and their negative effects can be circumvented so that the crucibles remain available and useful for any processing requirement. The purpose of the present volume is to address this area.

Even though metal–crucible interactions exist technically at all temperatures and in all states of aggregation, they are important and warrant remedial action particularly when the temperatures are high and melts are involved. These are exactly the conditions when thermodynamics expands the possibilities for interactions, and kinetics causes them actually to occur. Systems with the melts and containers in contact at high temperatures are encountered in two major areas of technology. One is in-process metallurgy and post-reduction processing of materials, and the other is in liquid metal/alloy cooling loops. Both are important areas of technology and have attracted considerable research effort all over the world. There are, however, major differences in the problems faced in each of these areas and the way solutions to the problems are developed. They have more differences than commonalities. To maintain focus, this book covers only the processing-oriented issues of metal–crucible interactions.

It would not be accurate to conclude that the size of the crucible does or does not determine the number, type and extent of interactions, but there is a loose linkage. A larger system tends to be more complex. It would therefore be unrealistic to isolate and consider only interaction between the metal and its container in such cases. Attention needs to be paid to all major reactions and processes. Often, the components of the melt attack the crucible synergistically. A great example is the corrosive attack in the belly band zone of aluminium melting and holding furnaces. Generally, it is the slag rather than the metal that attacks the refractory, so much so that an oxidized metal melt becomes more corrosive then when it was purer. Sometimes, a minor alloy addition to the metal or a binder in the crucible tends to become the source of interaction.

In this chapter, the history of metal processing was considered from the viewpoint of crucibles used at various time periods. The nature, type and role of a crucible depend on the format of the overall equipment, and an attempt was made to provide a glimpse of the major equipment and processes that have been used in metal processing over the years. The objective is to provide a framework for considering the many types of possible metal–crucible interactions in the next three chapters. In the following pages, an attempt has been made to present the various facets of this interesting issue, which has challenged human ingenuity throughout known history and continues to do so even today.

Each of the few metal–crucible interactions selected for specific coverage in this book is representative of the type of problems faced by many more metal–crucible systems. The drivers for interactions and the remedial actions that can circumvent or offset the damages are described. Most of the issues concerning any metal–crucible pair can be resolved by analogy to those covered specifically here.

A collection of metal–crucible pairs that have, at one time or other, been found workable is listed in Table 1.1.

Appearing next is Table 1.2, which lists various crucibles that have been recommended for use in the evaporation of metals.

The entries in this table have been compiled from published information from equipment manufacturers. Evaporation does not necessarily involve melting, and as the evaporation is carried out in the context of thin film applications, the duration of contact between the metal and the crucible is expected to be short and the quantity involved smaller. The tables are useful for identifying the members of the pair

TABLE 1.1
Composition of Crucibles Used for Melting Metals

Metal	Crucible Materials (Remarks)	References
Alkali metals	Refractory sulphides, Single crystal MgO (The refractory sulphides include BaS, CeS, ThS, Ce_3S_4, Ce_2S_3, Th_2S_3, Th_4S_7 and ThS_2)	Eastman et al. 1951, Brewer et al. 1948
Alkaline earth metals	Refractory sulphides	Eastman et al. 1951, Brewer et al. 1948
Aluminium	Zircon, Zirconia, Graphite, Clay-graphite, Refractory sulphides, Fire clay (dense), Aluminium nitride (There is least contamination with zircon)	Eastman et al. 1951, Brewer et al. 1948, Knauft 1943, Long and Foster 1959
Beryllium	Beryllium oxide, Graphite, Refractory sulphides (Melting must be conducted in a vacuum)	Kaufmann and Gordon 1947
Bismuth	Cerium sulphide, Thorium sulphide	Eastman et al. 1951, Brewer et al. 1948
Boron	Boron nitride	Finlay et al. 1952
Brass	Magnesium oxide, Graphite, Clay-graphite High-grade fire clay (Charcoal or fluxing material on surface of melt helps to reduce oxidation)	Harvey 1945
Cerium	Titanium nitride, Zirconium nitride, Cerium sulphide, Barium sulphide (Cerium sulphide crucibles resist attack better if the porosity[a] is from 10 to 15%)	Eastman et al. 1951, Brewer et al. 1948
Chromium	Aluminium oxide	Gayler 1930
Copper	Silica, Magnesium oxide, Aluminium oxide Clay-graphite	Harvey 1945, Booth 1940
Gold	Magnesium oxide, Aluminium oxide	Taylor et al. 1948
Halide metals	Refractory sulphides	Eastman et al. 1951, Brewer et al.1948
Iridium	Thorium oxide	Taylor et al. 1948
Iron	Magnesium oxide, Beryllium oxide, Aluminium oxide, Zircon, Fire clay, Thorium oxide, Refractory sulphides (For extremely pure iron, beryllia or thoria is best)	Eastman et al. 1951, Brewer et al. 1948, Richardson 1935, Turner 1931, Thompson and Cleaves 1939
Lead	Iron, Chromite, Magnesium oxide, Fire clay (dense)	Harvey 1945, Thews 1931
Magnesium	Steel plate coated with Aluminium, Heat-coat mixture of 10 to 60% ZrO_2, 7 to 25% MgO, 30 to 80% Al_2O_3, Cerium sulphide, Thorium sulphide (The steel plate should be sand-blasted before spraying and heat-treated after spraying)	Brewer et al. 1948, Taylor et al. 1948, Field 1947
Manganese	Aluminium oxide, Magnesium oxide, Thorium oxide (The metal should be pure, since a manganese-rich alloy will pick up silica)	Gayler 1930

(Continued)

TABLE 1.1 (CONTINUED)
Composition of Crucibles Used for Melting Metals

Metal	Crucible Materials (Remarks)	References
Nickel	Zircon, Magnesium oxide, Mullite, Fire clay (dense) (For melting pure nickel, zircon or magnesia in a vacuum is best)	Seybolt 1946, Jordan and Swanger 1930
Platinum	Thorium oxide, Zirconium oxide, Magnesium oxide, Zircon (Also for platinum-rich alloys)	Taylor et al. 1948, Baldwin1948
Rareearth metals	Refractory sulphides, Tantalum, Tungsten	Eastman et al. 1951, Brewer et al. 1948, Dennison et al. 1966a, 1966b
Rhodium	Zirconium oxide	Taylor et al. 1948
Silicon	Silica	Gayler 1930
Silver	Graphite, Magnesium oxide, Aluminium oxide	Taylor et al. 1948
Steel	Magnesium oxide, Silica (Basic steels are melted in magnesia, and acidic steels are melted in silica)	Taylor et al. 1948
Thorium	Cerium sulphide, Thorium sulphide, Titanium nitride, Zirconium nitride (The metal sticks to the sulphide crucibles)	Brewer et al. 1948, Richardson 1935
Tin	Magnesium oxide, Cerium sulphide, Thorium sulphide, Fire clay	Eastman et al. 1951, Brewer et al. 1948
Titanium	Cerium sulphide, Barium sulphide, Thorium sulphide (Titanium sticks to the sulphide crucibles. There is slight attack of the thoria, melting is accomplished in either a vacuum or an argon atmosphere)	Eastman et al. 1951, Brewer et al. 1948, Brace 1948
Uranium	Magnesium oxide, Calcium oxide, Cerium sulphide, Titanium nitride, Zirconium nitride, Thorium oxide, Thorium sulphide, Graphite, Nickel	Brewer et al. 1948, Richardson 1935
Vanadium	Beryllium oxide, Magnesium oxide, Calcium oxide, Titanium nitride (These materials were calculated to be suitable for containing molten vanadium)	
Zinc	Flint fire clay, Cerium sulphide, Thorium sulphide, Clay-graphite	Eastman et al. 1951
Zirconium	Zirconium dioxide, Thorium oxide	Richardson 1935

Source: Schwartz, M.A., White, G.D., and Curtis, C.E., *Crucible Handbook: A Compilation of Data on Crucibles Used for Calcining, Sintering, Melting, and Casting, United States Atomic Energy Commission.* ORNL-1354, Oak Ridge National Laboratory, Oak Ridge, TN, 1953.

[a] Porosity is defined as the volume of airspace expressed as percentage of total volume of the piece.

(metal–crucible) for further investigation. Like the Ellingham diagrams given in the next chapter, the information provided by the table should at best be regarded as a starting point for considering the combination, and good starting points are definitely important. As will be seen, fairly repeatedly, in the rest of the book, metal–crucible interactions, both the extent and the rate, are determined by multiple factors:

TABLE 1.2
Crucibles Suitable for Evaporation of Metals

Metal	Melting point, °C	Temperature, °C[a]	Crucible materials (Comments)
Aluminium	660	972	AlN, TiB_2-BN, ZrB_2, BN, Vitreous carbon, W
Antimony	630	425	BN, C, Al_2O_3, Alumina-coated Mo and Ta
Arsenic	817	210	Al_2O_3, BeO, Vitreous carbon
Barium	729	462	W, Ta, Mo (Wets refractory metals without alloying. Reacts with ceramics)
Beryllium	1289	987	BeO, Graphite, Vitreous carbon W, Ta (Wets W, Ta)
Bismuth	271	520	Al_2O_3, Vitreous carbon, W, Mo, Ta
Boron	2092	1797	Graphite, Vitreous carbon (Forms carbide with the container)
Cadmium	321	180	Al_2O_3, Quartz, W, Mo, Ta
Calcium	842	459	Al_2O_3, Quartz, W
Carbon	3826	2107	
Cerium	798	1380	Al_2O_3, BeO, Vitreous carbon, W, Ta
Cesium	28	75	Quartz, Stainless steel
Chromium	1863	1162	Vitreous carbon
Cobalt	1495	1262	Al_2O_3, BeO, W, Nb (Alloys with refractory metals)
Copper	1085	1017	Al_2O_3, Mo, Ta
Dysprosium	1412	900	Ta
Erbium	1529	930	W, Ta
Europium	822	460	Al_2O_3, W, Ta
Gadolinium	1313	1175	Al_2O_3, Ta
Gallium	30	907	AlN, Al_2O_3, BeO, Quartz (Alloys with refractory metals, attacks crucibles above 1000°C)
Germanium	937	1142	Quartz, Al_2O_3, W, Ta, Carbon (Wets W, Ta, Mo)
Gold	1064	1132	Al_2O_3, BN, W, Vitreous carbon (Wets W)
Hafnium	2231	3090	
Holmium	1474	950	W
Indium	157	742	Graphite, Al_2O_3, W, Mo (Wets tungsten and copper)
Iridium	2447	2380	ThO_2
Iron	1538	1207	Al_2O_3, BeO, W (Attacks tungsten)
Lanthanum	921	1388	Al_2O_3, W, Ta
Lead	328	497	Al_2O_3, Quartz, W, Mo, Ta
Lithium	181	407	Al_2O_3, BeO, Ta, Stainless steel
Lutetium	1663	1300	Al_2O_3, Ta
Magnesium	649	327	Al_2O_3, Vitreous carbon, W, Mo, Ta, Nb
Manganese	1244	647	Al_2O_3, BeO, W, Ta, Mo (Wets refractory metals)
Mercury	−39	−6	
Molybdenum	2623	2117	
Neodymium	1021	1062	Al_2O_3, Ta (Low solubility of tantalum)

(Continued)

TABLE 1.2 (CONTINUED)
Crucibles Suitable for Evaporation of Metals

Metal	Melting point, °C	Temperature, °C[a]	Crucible materials (Comments)
Nickel	1455	1262	Al_2O_3, BeO, Vitreous carbon, W (Alloys with refractory metals)
Niobium	2468	2287	W
Osmium	3033	2760	
Palladium	1554	1192	Al_2O_3, BeO, W (Alloys with refractory metals)
Platinum	1769	1747	Al_2O_3, ThO_2, Graphite, W (Alloys with metals)
Plutonium	641		W
Polonium	254	244	Quartz
Potassium	63	125	Quartz, Mo
Radium	700	416	
Rhenium	3186	2571	
Rhodium	1963	1707	W, ThO_2, Vitreous carbon
Rubidium	39	111	Quartz
Ruthenium	2334	2260	W
Samarium	1074	573	Al_2O_3, Ta
Scandium	1541	1100	Al_2O_3, BeO, W (Alloys with tantalum)
Selenium	221	170	Al_2O_3, Vitreous carbon, W, Mo
Silicon	1414	1337	BeO, Ta, Vitreous carbon, W (alloys with W)
Silver	962	832	Al_2O_3, W, Mo
Sodium	98	192	Quartz, Tantalum, Stainless steel
Strontium	769	403	W, Ta, Mo, Vitreous carbon (Wets refractory metals without alloying)
Tantalum	3020	2590	
Tellurium	452	277	Al_2O_3, Quartz, W, Ta (Wets refractory metals without alloying)
Terbium	1356	1150	Al_2O_3, Ta
Thallium	304	470	Al_2O_3, Quartz, W, Ta
Thorium	1755	1925	W, Ta, Mo (Wets W)
Thulium	1545	680	Al_2O_3, Ta
Tin	232	997	Al_2O_3, Mo, Ta (Wets molybdenum)
Titanium	1670	1453	TiC, Vitreous carbon, W (Alloys with refractory metals)
Tungsten	3422	2757	
Uranium	1135	1582	Y_2O_3, Mo, W
Vanadium	1910	1547	W, Mo (Wets molybdenum. Alloys slightly with W)
Ytterbium	819	417	Ta
Yttrium	1522	1157	Al_2O_3, W, Ta (High tantalum solubility)
Zinc	420	250	Al_2O_3, Quartz, Mo, W, Ta (Wets refractory metals)
Zirconium	1855	1987	W

[a] Temperature in°C for 0.01 Pa vapour pressure.

temperature for one, duration of contact, identity of the melt and its impurity content, viscosity of the melt, its actual wetting characteristic, impurities in the refractory, microstructure of the refractory, its porosity, and pore sizes and their distribution, to name the most important. Even when we consider a metallic crucible, sometimes whether it is a single crystal or polycrystalline determines whether it would actually be useful as a crucible. The original papers that relate to the table entries usually answer many of these. In the end, the value of an actual validation experiment cannot be overemphasized.

2 Basics of Interactions

2.1 INTRODUCTION

The tendency of materials to react with one another can be generally predicted by considering their thermodynamic properties. When the possibility of interactions at elevated temperatures is considered, kinetics is fast enough to render thermodynamic predictions closely reflect real situations. Graphical representation of the thermodynamics of the system as Ellingham and phase diagrams is especially valuable to reveal possible interactions. Even though in many instances, more than two components are involved in the interaction, and multi-component phase diagrams must be consulted when accessible, much useful information can still be obtained from the relevant binary phase diagrams that may be more readily available. A variety of transport processes contribute to the materialization of metal–crucible interaction. Apart from thermodynamic properties, there are requirements placed on the crucible materials as regards their thermo-physical properties, both intrinsic and vis-à-vis the material being processed. Fabricability of the crucible in the required form is another key consideration.

Chemical analysis is, obviously, very important in assessing the interaction. This, in addition to X-ray, microscopic and microprobe analyses of the interfaces, helps to map the nature and extent of interaction more completely. These tools and techniques are all relevant in the context of predicting, explaining and thus managing metal–crucible interactions.

Thermodynamic considerations are the usual and also the right way to shortlist materials that could serve as crucibles for processing melts. Management of interactions involves attention to several other factors, including properties and processes that act together. Wetting and penetration of melt in refractory pores is a common issue. Volume change during the process occurs due to formation of the product by chemical interaction or due to temperature changes and the expansion characteristics of the various phases present in the system. Volume change usually aggravates and occasionally prevents continuation of melt ingestion. Thermal shock and its consequences are a major issue causing or aggravating degradation. As emphasized in Chapter 1, the scope of metal–crucible interactions covered in this work is fairly broad, relating principally to production and processing of materials and not to usages like liquid metal loops in energy systems. The term *crucible* is used to denote essentially all surfaces that contact the process liquids. The origin of the liquid may be direct or collateral. Interactions of vapours and gases are considered only to the extent that they support or suppress the liquid–solid interactions and act concurrently.

2.2 THERMODYNAMICS

Materials in contact tend to interact with one another. The interaction is greater and faster at high temperatures and when one of the materials is a fluid. Such interactions

DOI: 10.1201/9780429345562-2

may occur on purpose, being a part of process being implemented, in which case conditions are deliberately created to facilitate its occurrence. The other group of interactions occur not on purpose and in spite of not being intended either, e.g. metal–crucible interactions. Such interactions are seldom beneficial, and efforts are always made to avoid them or at least, control them to the least harmful levels.

Metal–crucible interactions are driven by thermodynamics, whether it results in the formation of a simple Raoultian solution between the melt and the crucible or in an all-out chemical reaction resulting in products that have little resemblance to the original melt or the crucible. Checking on the thermodynamic properties of the system is a necessary and proper first step in assessing these interactions.

Basically, the question that needs to be answered here, as in every area of materials processing, pertains to whether a compound MX_n, where M is a metal and X is a non-metal, will react with an element R, another metal, at a given temperature. A special case is when the crucible and the melt are both elements, e.g. carbon and iron, or tungsten and lanthanum. The situation is simpler here but a special case of the more general scenario presently considered. In the context of metal–crucible interactions, MX_n may be considered a ceramic compound and R the metal to be contained by the ceramic. The question is answered conclusively by thermodynamics, but the answer is accessible only when the thermodynamic properties of all the entities in the reaction are known or can be estimated. The present question may be represented by the general reaction:

$$MX_n + iR = M + iRX_{n/i}$$ (2.1)

where M is a metal; X is oxygen, carbon, nitrogen, boron, sulphur; and R may be a metal such as titanium, vanadium, aluminium or any element. The reaction is possible only when at a chosen reaction temperature, the difference between the free energies of formation, ΔG, of the compound formed by the interacting element, $RX_{n/i}$ and the starting compound, MX_n, is negative: in other words, if the condition

$$i\Delta G\left(RX_{n/i}\right) - \Delta G\left(MX_n\right) < 0$$ (2.2)

holds good. Therefore, the free energies of formation of the compounds (oxides, nitrides and so on) of all the elements and their dependence on temperature determine which element R will react with MX_n to release the M. To make this information easy to assess, the data on the free energies of formation of compounds were presented graphically as straight line plots with temperature as abscissa and standard free energy of formation as ordinate, first by Ellingham (1944) and later by Richardson (1948) and Richardson et al. (1950). The Ellingham diagrams of compounds relevant to the selection of crucible materials are given in Figures 2.1 to 2.5 (oxides, carbides, nitrides, sulphides, fluorides). Data from the U.S. Bureau of Mines Publications (Pankratz et al. 1984) have been used in preparing these diagrams.

The metals titanium, zirconium, vanadium and every one of the 16 naturally occurring rare earths are extremely reactive in the sense that they tend to interact strongly with all usual crucible materials and also with some unusual materials that

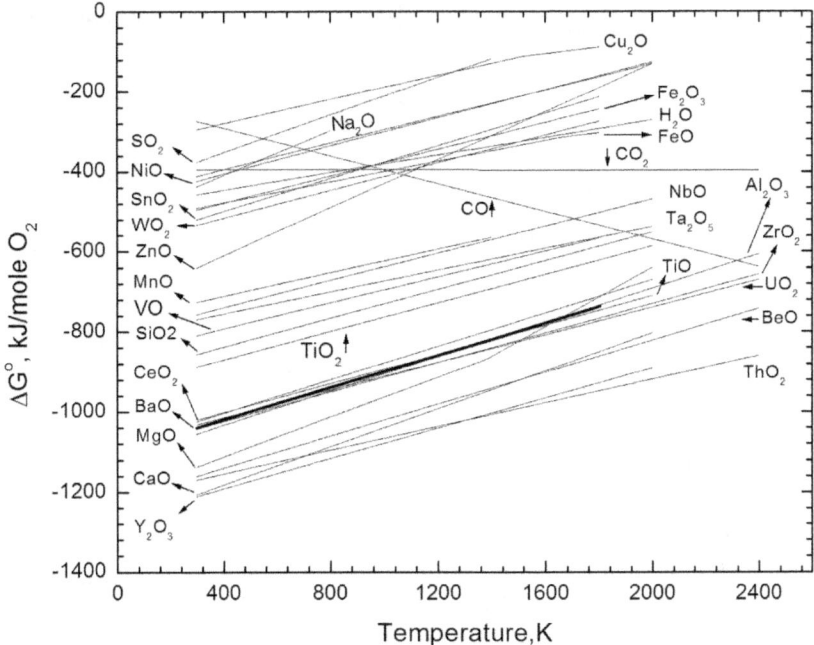

FIGURE 2.1   Ellingham diagram of oxides.

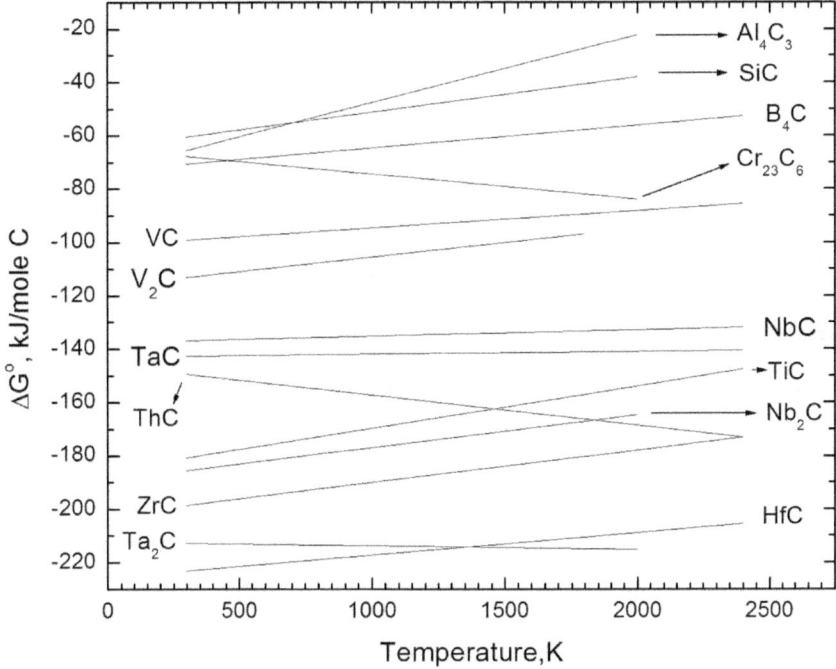

FIGURE 2.2   Ellingham diagram of carbides.

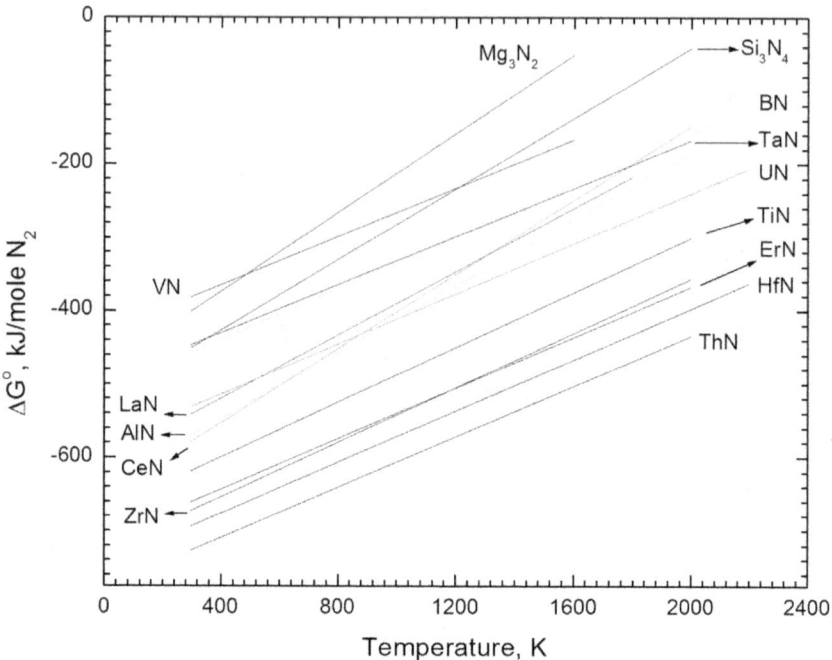

FIGURE 2.3   Ellingham diagram of nitrides.

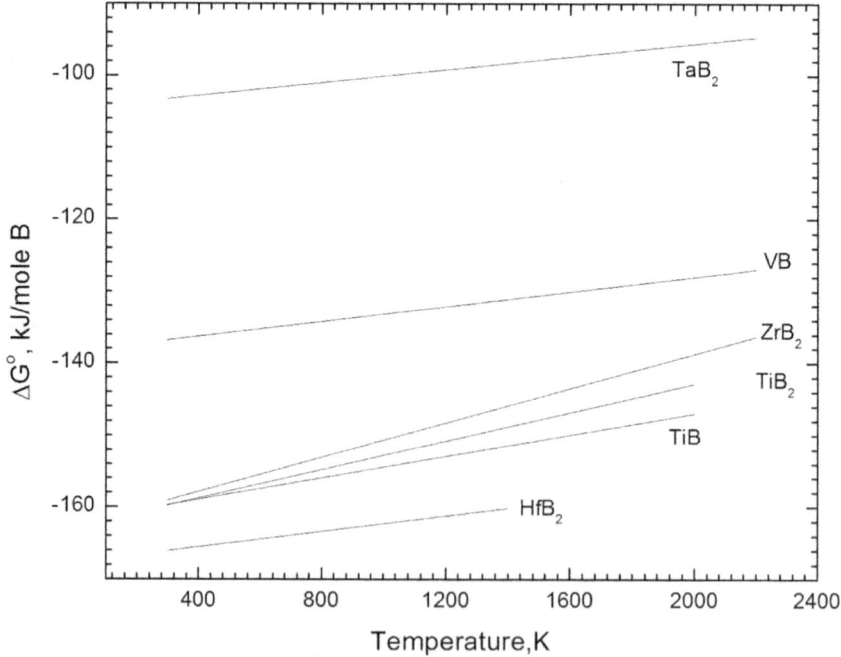

FIGURE 2.4   Ellingham diagram of borides.

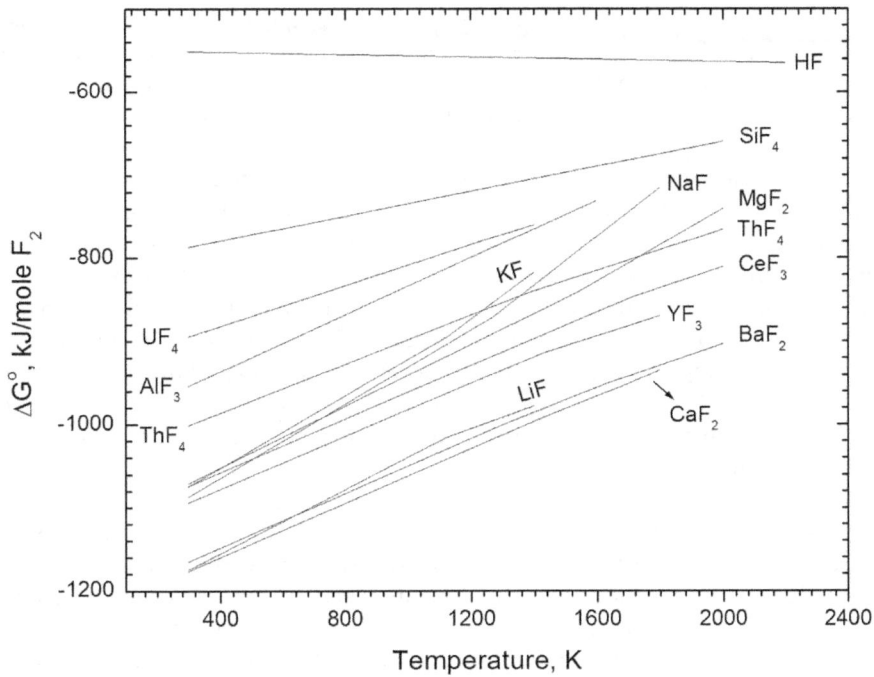

FIGURE 2.5   Ellingham diagram of fluorides.

are considered to be rather stable. It seems instructive, therefore, to look in detail at the process of selecting crucibles for them, e.g. for titanium. For instance, if it is planned to use an oxide crucible, say alumina, the free energies of formation of titanium oxide will be compared with the free energy of formation of alumina. Since titanium forms many stable oxides, for comparison, the most stable oxide phase of titanium, TiO, will have to be selected. The free energy of formation of Al_2O_3 will be compared with the free energy of formation of TiO. The comparison must be carried out on an equivalent basis; that is, the molar free energy of formation of alumina represents the free energy of formation of alumina when it is formed with the consumption of 3 gram atoms or 1.5 gram moles of oxygen. The molar free energy of formation of alumina when multiplied by the factor 2/3yields the value of free energy formation of alumina per gram mole of oxygen. Similarly, the molar free energy of formation of TiO represents the free energy of formation of TiO when it is formed with the consumption of 1 gram atom or 0.5 gram moles of oxygen. The molar free energy of formation of TiO when multiplied by a factor of two yields the free energy of formation of TiO per gram mole of oxygen. So, before comparing the values of free energy of formation of oxides, the values for free energy of formation per mole of the oxide should be converted to the free energy of formation per mole of oxygen by multiplying with the appropriate factor, and the free energies of formation of oxides per mole of oxygen must be compared. If the formula of the compound for which the molar free energy values are listed is MxNy, then the factor is (2/y), y being the subscript of the anion. It may be oxygen, nitrogen, fluorine, carbon, silicon, sulphur or phosphorus. For gases, the mole or 1 gram mole equals 2 gram atoms, and for all other elements mentioned, 1 gram mole equals 1 gram atom. This comparison of free energy formation must be done only in this way irrespective of the compound that is considered, on an equivalent basis. For example, if we are considering $BaZrO_3$ as the crucible, then the factor is still 2/y (i.e. 2/3).

On comparison of the free energies of formation of the refractory compound with that of the corresponding titanium compound, on an equivalent basis, as explained in the previous paragraph, if it is found that the free energy of formation of the refractory compound has a significantly larger negative value than the free energy of formation of the titanium compound, the refractory compound can be used as a crucible for containing titanium as a first approximation. If the free energy of formation of the refractory compound has a smaller negative value that that of the titanium compound, the crucible will be useless for containing titanium. If the values are more or less similar, further evaluations (calculations and experiments) will be necessary to comment on the suitability of the crucible.

A plot of the standard free energy of formation of a compound as a function of temperature is called an Ellingham line. In a single plot, several Ellingham lines can be drawn, all for the same type of compound (an oxide or a carbide), for a number of metals. The ordinate is the free energy formation per mole of oxygen or free energy of formation per gram atom of carbon, and so on. The Ellingham diagrams for a number of compounds are given in Figures 2.1 to 2.5. In the way they are plotted, these diagrams directly depict the relative stability of compounds of the same type. Elements with their Ellingham lines towards the bottom of the figure are more stable than those whose lines appear towards the top of the figure. It is only necessary that the Ellingham lines of the most stable compound of the metal to be melted and of the compound for the crucible material are both plotted there. The bottom most line is that of the compound that is the most stable among all the compounds appearing in the figure. All it takes is a look at these figures to know which refractory can be a potential crucible for a given metal. The larger the vertical gap between the lines, the greater is the relative stability between the compounds concerned.

The free energy of formation mentioned so far is actually the standard free energy of formation. It gives the free energy values corresponding to a process where the reactants and products are in the standard states. The standard state simply refers to the stable state at the given temperature and 1 atmosphere (ambient) pressure. Consider the standard free energy of formation of zirconium oxide, ZrO_2. The value refers to the free energy change when 1 gram atom of zirconium at an ambient pressure of 1 atmosphere combines with 1 gram mole of oxygen gas at 1 atmosphere pressure to yield 1 gram mole of ZrO_2 at an ambient pressure of 1 atmosphere, all at a fixed given temperature. The idealized relative stability information provided by the Ellingham diagram should only be considered as the starting point in the evaluation of relative stabilities of metals and compounds in real-life situations.

If, for example, the group of oxides Al_2O_3, BeO, CaO, MgO, SiO_2 and Y_2O_3 is considered, then under standard conditions (i.e. 1 atmosphere pressure at the given temperature), the order of stability, in descending order, among the oxides is (most stable) $CaO > RE_2O_3 > MgO > Al_2O_3 >> SiO_2$ (least stable). It is implicit in the statements made earlier that each of the solid phases is pure and the pressure of the gas phase is 1 atmosphere. Similar plots can be made and relative stabilities presented for any type of compound and for any number of metals in a group, provided thermodynamic data is available.

The picture under non-standard conditions that, more often than not, more closely approximates the real process condition is somewhat different. Here, the reaction mixture need not be composed of only pure constituents R, M and $RX_{n/i}$, and the ambient pressure may not be 1 atmosphere. In other words, the activity, a, of each of the constituents, a_R, etc., need not be kept equal to 1. Consider the following thermodynamic expressions.

$$\Delta G = \Delta G^0 + RT\ln Q \tag{2.3}$$

$$Q = a(M)a^i(RX_{n/i})/a(MX_n)a^i(R) \tag{2.4}$$

Even when ΔG^0 is positive, depending on the value of Q, ΔG can become negative, and reactions can occur when none would be expected considering only standard free energies. Usually, solutions form, and the activity of components appearing in the numerator of the expression for Q becomes low, making Q less than 1. There is a convenient way to obtain a small value for Q in reactions where the product is a gas or a vapour. The activity term is replaced by partial pressure, and by decreasing the partial pressure of the product (gas), a low value for Q is generated. Depending on how small (less than 1) the value of Q is and how high the temperature is, the term $RT\ln Q$ can become a large negative value, and a strong possibility of reaction emerges. It will be seen later that this situation develops when calcia or baria is used as a crucible at high temperatures and in vacuum.

The extent of reaction of metals with oxides is important for the problem of finding suitable refractories for the melting of metals and alloys. It was emphasized by Kubaschewski and Alcock (1979) that the higher the temperature, the less likely are the kinetic checks, and the more significant are the positions of thermochemical equilibria.

2.3   THE TITANIUM–OXYGEN SYSTEM

Even though Ellingham diagrams are available or can be created for any compound type, the most frequently used is the one for the oxides. Refractory oxides are the most widely used crucible materials, and oxygen contamination is among the most prevalent and also the most critical of the consequences of metal–crucible interaction. An examination of the Ellingham diagram of oxides shows up a fairly large number of choices for crucible materials for titanium. The case of titanium is exemplary in many ways. The melting point of the metal, its other physical properties, its chemical reactivity and alloying behaviour, taken together, bring to the fore most of the problems that can come up in any metal–crucible interaction. It is a suitable case to elaborate on.

It is seen that CaO and Y_2O_3 are potential crucible materials for direct melting of Ti because their free energies are more negative at all temperatures, whereas SiO_2 and Al_2O_3 have a ΔG^0_f (kJ/mole of oxygen) that is less negative than that of TiO (Lochbichlerand Friedrich 2007; Kostov and Friedrich 2006; Kostov and Friedrich 2005) at $T \geq 1600$ K. The latter group of oxides cannot, therefore, be used as materials for melting crucibles. The value of the free energy of formation ZrO_2 is close to the one for TiO, and an offhand comment would not be appropriate.

The simplified thermodynamic approach, which only considers the stability of the crucible material vis-à-vis the most stable oxide of the metal to be melted (presently TiO), overlooks certain key realities of the interaction of the melt with the crucible material. The phase diagram, Figure 2.6, of the oxygen–titanium system (Massalski 1990b) reveals that oxygen has a large solubility in Ti, and the conclusions derived from the Ellingham diagram refer to a titanium that is in fact a saturated solution of oxygen in titanium (i.e. titanium loaded with up to approximately 30 atom percent oxygen in solid solution). A similar situation prevails in many other systems also, e.g. O–Zr (Massalski 1990b).

To address this and related issues where a simple comparison of free energies will not present a complete or sufficiently accurate picture of possible interactions, there are very useful software packages that perform detailed thermodynamic calculations and arrive at the equilibrium composition of a given system by minimizing the total

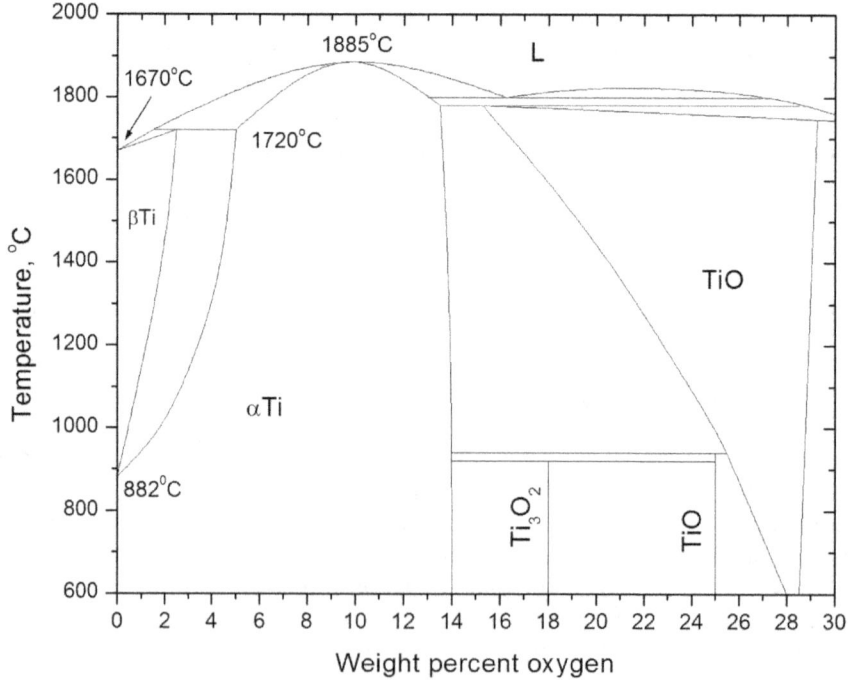

FIGURE 2.6   Titanium–oxygen phase diagram (From Massalski, T.B. 1990b).

free energy of the system under specified conditions. The conclusions are limited only by the reliability and extent of input information on the various thermodynamic data available for the system being assessed. Fact Sage has been a popular package. In this context, an assessment of the interaction thermodynamics between TiAlV melts and a number of binary and mixed oxides as candidate crucible materials was carried outby Kostov and Friedrich (2005, 2006).

It was concluded that out of six binary oxides (CaO, Y_2O_3, ZrO_2, MgO, Al_2O_3 and SiO_2) and many of their combinations (oxide spinels), only CaO, Y_2O_3, ZrO_2 and Al_2O_3 have positive values of free energy, ΔG_r^0, in respect to their interaction with TiAlV melts at T = 1273–1973 K in the whole range of Ti concentrations. However, for the alloys with high Ti concentration and high melting temperature, ΔG_r^0 with Al_2O_3 is close to zero, which eliminates its use as a crucible.

While this approach is highly useful and practical for shortlisting potential crucible materials, it is also instructive to follow the mechanics of evaluation of potential crucible materials as outlined by Lyon et al (1973) and later by Okabe et al. 1991.

The selection of a crucible for melting titanium remains, as of now, a work in progress. The reason for this state of affairs is more readily understood if the attempts already made at a solution are revisited.

The free energies of formation of refractory compounds are compared with those of the corresponding titanium phases (TiO, TiN and TiC). Among the oxide ceramics, Al_2O_3, ZrO_2, ThO_2, MgO, CaO, BeO andY_2O_3 are good for possible experimental evaluation. This approach is, as indicated earlier, incomplete. It overlooks the need to consider solution effects, i.e. the complications introduced due to titanium metal accommodating considerable oxygen in solid solution. The picture become clearer if the partial Gibbs energy diagram for the titanium–oxygen system given in Kubaschewski and Alcock (1979) is considered. The partial Gibbs energy of dissociation of 1 mole of oxygen over the entire composition range of the Ti-O binary is given. The value, which was approximately 750 kJ in the TiO + α two-phase region, rises rather steeply as the system enters the α solid solution range (at about 15 wt% oxygen at 1000°C) and flattens (at 940 kJ) briefly in the β+α two-phase region (at around 2 wt% oxygen) followed by another extremely steep rise in the β Ti–O solid solution region to 1000 kJ. This information indicates that the least stable oxide crucible that can still be used for melting titanium would be calcia. In going by the simple Ellingham plots to assess the reducibility of oxides, the stability of the titanium–oxygen system to be tackled was underestimated by nearly 250 kJ.

As titanium has not only great affinity but also large solubility for interstitials, the possibility of occurrence of a reaction with a crucible needs to be evaluated by considering the partial molar free energies of solution for interstitials in titanium. For example, at 1400°C, the free energy of formation of ZrO_2 is more negative than any titanium oxide phase, but comparison of this with a free energy of formation of titanium–10 atomic percent oxygen solution indicates existence of a driving force for dissolution of the zirconia (Lyon et al. 1973; Gingerich 1968) into titanium. It turns out that oxides such as Al_2O_3 and SiO_2 are not compatible with (cannot be used to melt) titanium, even at relatively low temperatures. For it to be considered as a

TABLE 2.1

Activity of Titanium at the Solidus Temperatures of Various Titanium Alloys

Alloy	Activity of titanium	Solidus temperature, °C
Ti	1.00	1668
Ti6Al4V(wt.%)	0.87	1686
Fe70Ti(wt.%)	0.56	1109
Ti44.5Al6Nb0.2C0.2B (at .%)	0.29	1582
Ti45Al7Nb1Mo0.2B (at .%)	0.28	1585
Ti47Al2Cr0.2Si (at .%)	0.27	1528
Ti50Al (at .%)	0.24	1517
Fe30Ti (wt .%)	0.07	1433

Source: Friedrich et al., 2016.

potential crucible material for melting titanium, the prospective oxide should be able to coexist with a very dilute titanium–oxygen solution at temperatures near the melting point of titanium. This requirement eliminates most refractory oxides with the exception of CaO, ThO_2, MgO, the trivalent rare earths, and BeO among the binary oxides. It is entirely likely that if the titanium alloys are considered, the partial molar free energy of oxygen in the alloy at comparable concentrations is less negative than the binary Ti–O solution, making the choice of crucibles somewhat wider. This is because the activity of titanium is less than 1. That makes Q, described earlier, a larger value, and ΔG becomes less negative. Friedrich et al. 2016 have given the values for the activity of titanium in various popular titanium alloys. The data presented by them in the form of a figure is listed in Table 2.1.

In the TiAl group of alloys containing approximately 50 atom percent titanium, the activity of titanium is much smaller than 0.5. It is approximately 0.3 to 0.25 and does not change much with other alloying components such as niobium, molybdenum, silicon and boron. The trend is the same with the FeTi alloys. The titanium activity decreases to 0.56 with the mole fraction of titanium in the alloy coming down to 0.73. There is a rather drastic decrease in titanium activity to 0.07 when the mole fraction of titanium in the FeTi alloy is 0.33. Incidentally, all this is good news when it comes to finding a crucible for melting titanium alloys, because much less aggression can be expected from these alloys than from pure titanium.

There is considerable solubility of oxygen in titanium (Figure 2.6). Yttria has an oxygen stoichiometry range and can exist as an oxygen-deficient compound (Figure 2.7). When evaluating the possibility of melting titanium in an yttria crucible, the reaction considered by Lyon et al. (1973) was, therefore,

$$Ti + Y_2O_3 = Ti(O \text{ in soln}) + Y_2O_{3-x} \qquad (2.5)$$

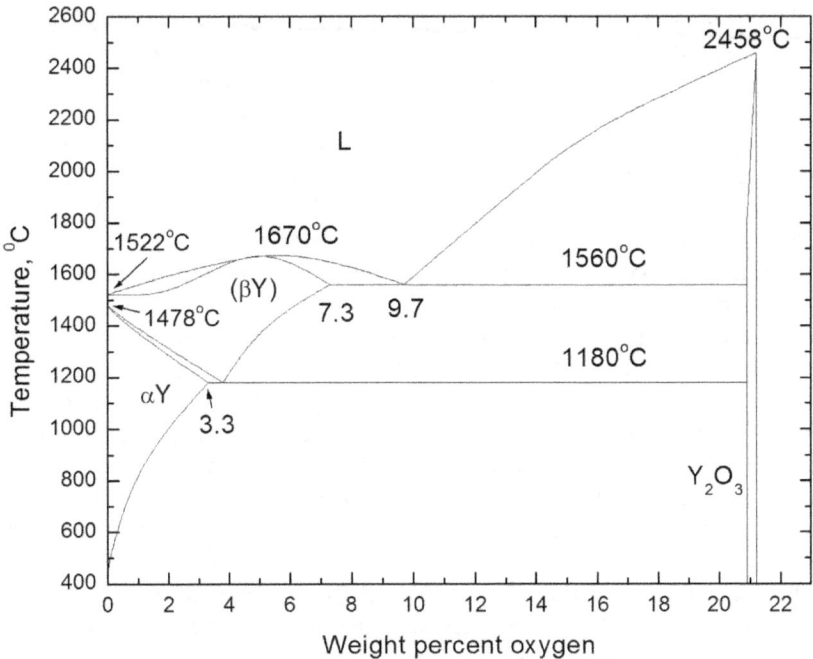

FIGURE 2.7   Yttrium–oxygen phase diagram (From Massalski, T.B. 1990b).

They determined the quantities necessary for evaluation of the free energy change for this equation in a series of mass spectrometric experiments performed on molten titanium in contact with yttria crucible liners. The purpose of the experiments was to determine the oxygen potential (partial pressure of oxygen over Ti–O solutions) in molten titanium of known oxygen content and the oxygen and yttrium potential of yttria of a known stoichiometry.

From the computed pressures over a range of oxygen concentrations, Lyon et al. (1973) determined the Sievert's law constant for oxygen in titanium at 1960 K as

$$X_O = 10^7 \, p_{O2}^{1/2} \tag{2.6}$$

At 1960 K, the oxygen partial pressure was 10^{-19} atm for oxygen-deficient yttria with an O/M of 1.47 (i.e. for $Y_2O_{2.94}$). Based on the observed partial pressure of yttrium over Y_2O_{3-X}, they determined that at 1960K, the yttria phase can exist at oxygen pressures down to 10^{-24} atm. At such low oxygen potential, the thermodynamic activity of Y approaches unity, and the titanium melt will pick up yttrium more than at any other composition. The yttrium content in titanium solution could go to possibly 20 weight percent when the activity of yttrium is about 0.25 in value. From mass spectrometric data, Lyon et al. computed the activity of yttrium to be 6×10^{-5} in $Y_2O_{2.94}$ at 1960 K. With this information, Lyon et al. (1973) determined the thermodynamic feasibility of producing an acceptable-quality titanium ingot in an yttria crucible.

If the objective is titanium containing 1500 ppm oxygen by weight, i.e. 4.5×10^{-3} atom fraction oxygen, the oxygen potential in molten titanium will need to be 2×10^{-19} atm p_{O2} at 1960 K. The value is obtained by substituting for X_O in the Sievert's relation. Stoichiometric Y_2O_3 at this temperature has a potential near 10^{-16} atm p_{O2}; hence, a reaction between stoichiometric Y_2O_3 and titanium is possible, and there will be oxygen ingestion from the yttria crucible into the titanium melt. Were the yttria crucible reduced to a composition of about $Y_2O_{2.94}$ prior to charging, then the oxygen potential would be matched between the oxide phase (10^{-19} atm at 1960K) and the titanium phase (2×10^{-19} atm), and driving force for transport of oxygen between them would be greatly decreased.

From an oxygen content standpoint, then, yttria with the lowest possible oxygen content (stoichiometry) would be the best. However, yttria has a rather narrow range of stoichiometry, and the O/M ratio cannot be any lower than 1.47 for yttria according to the diagram given in (Massalski 1990b). Incidentally, this diagram is redrawn from the one in Shunk (1969). However, the results of Lyon et al. (1973) show that the yttria phase can exist at oxygen pressures down to 10^{-24} atm at 1960 K, implying a possible lower oxygen stoichiometry limit for Y_2O_3, i.e. a value more than 0.06 for x in Y_2O_{3-x}.

Alluding to this possibility, Lyon et al. (1973) proceeded to observe that should titanium containing 1500 ppm oxygen be melted in a crucible whose stoichiometry is lower than $Y_2O_{2.94}$, then one would expect the titanium to give up some of its oxygen to the crucible. As regards yttrium pickup by titanium melt, Lyon et al. (1973) have given the following expression relating concentration and activity of yttrium in titanium at 1960 K:

$$a = 1.5 \times 10^{-2} w \tag{2.7}$$

where a is the thermodynamic activity of yttrium and w is the weight percent of yttrium expected to be dissolved in titanium. At 1960 K, if 400 ppm by weight is the maximum yttrium pick-up allowed, then the activity of yttrium must be 6×10^{-4} or less, and the oxygen pressure to which this yttrium activity corresponds is obtained from

$$\left(a_Y\right)\left(p_{O2}\right)^{0.75} = 3 \times 10^{-19} \tag{2.8}$$

where a_Y is the desired activity of yttrium and p_{O2} is the oxygen pressure in atmospheres. Solving this for oxygen gives $p_{O2} = 3 \times 10^{-21}$ atm. According to Eq. (2.6), the oxygen content corresponding to this partial pressure would be 5×10^{-4} atom fraction in the titanium. This value corresponds to less than 200 ppm by weight oxygen in titanium. It is therefore entirely feasible to expect good-quality titanium products (~200 ppm oxygen) melted in yttria crucibles, provided the yttria has been pre-treated to ensure that it is of the proper stoichiometry.

Using published (Kubaschewski and Dench 1954; Komarek and Silver 1962; Okabe et al. 1991) values for the partial molar free energies of Ti–O solutions and their variation

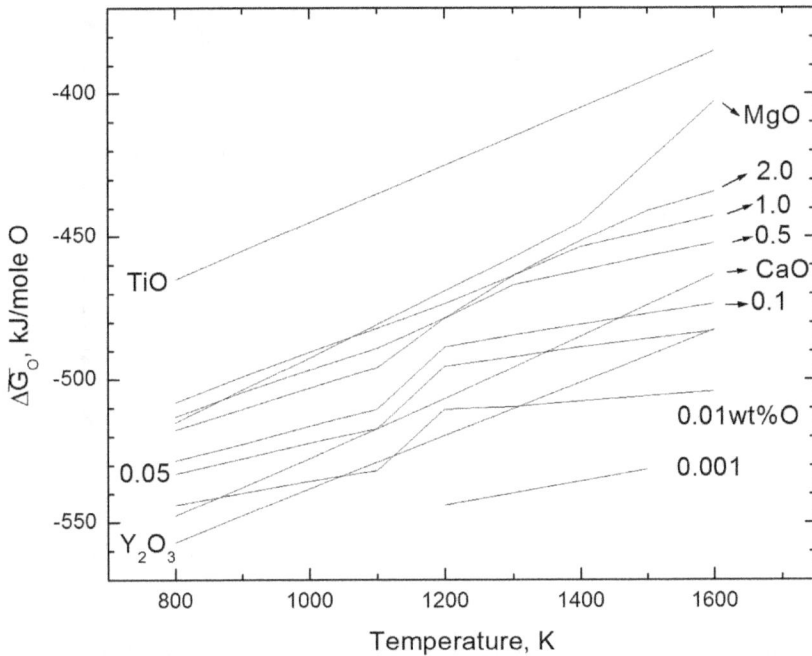

FIGURE 2.8 Oxygen potential of Ti–O solid solution and selected metal oxides (From Zhang et al. 2020).

with temperature, plots like the Ellingham diagrams can be generated. Zhang et al. (2020) presented in such a plot the temperature variation of equilibrium oxygen partial pressures over titanium–oxygen solid solutions at oxygen concentrations between 10 and 10,000 mass ppm, and the free energies of formation certain metal oxides. The plot given as Figure 2.8 effectively defines the limits of oxygen contamination that can be reached in titanium metal when using the pure stoichiometric oxides as crucibles. A magnesia crucible would result in oxygen contamination of at least 2 wt.%, a calcia crucible >0.1 wt% and yttria, as already mentioned, in the range of 0.05 wt%.

2.4   THE VANADIUM–OXYGEN SYSTEM

A situation similar to that for titanium exists in the vanadium–oxygen system too. There are multiple solid phases in the vanadium–oxygen system with varying stoichiometry ranges. The free energy diagram of the system in the composition range from V_2O_5 to pure vanadium is shown in Figures 2.9 and 2.10 (Allen et al. 1951). The vanadium–oxygen solid solution is less extensive than those of Ti–O or Zr–O. Vanadium metal dissolves a maximum of 0.25 wt.% oxygen at 1000°C and 5 wt% O at the eutectic temperature 1667°C. The stability of solution increases steeply as the oxygen content decreases; removal of oxygen (deoxidation) becomes more difficult as the metal becomes purer with respect to oxygen. In other words, as the purity of

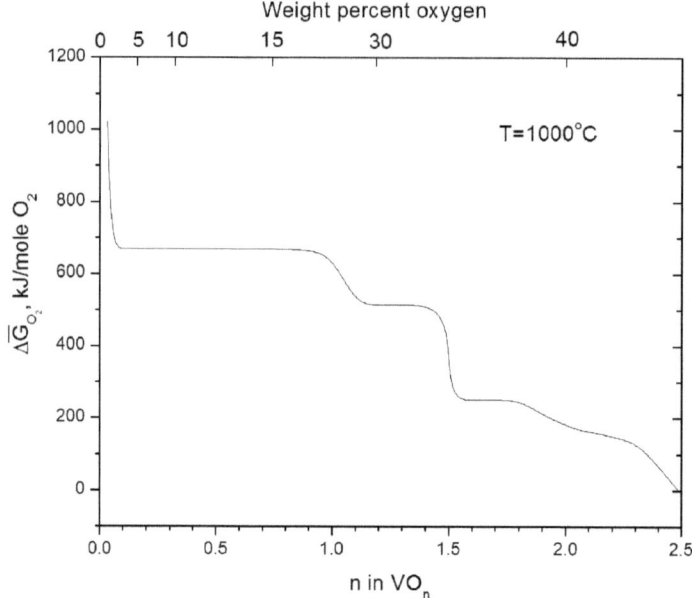

FIGURE 2.9 Partial molar free energy of oxygen in vanadium–oxygen system (From Allen, N.P., Kubaschewski, O., and Von Goldbeck, O. 1951).

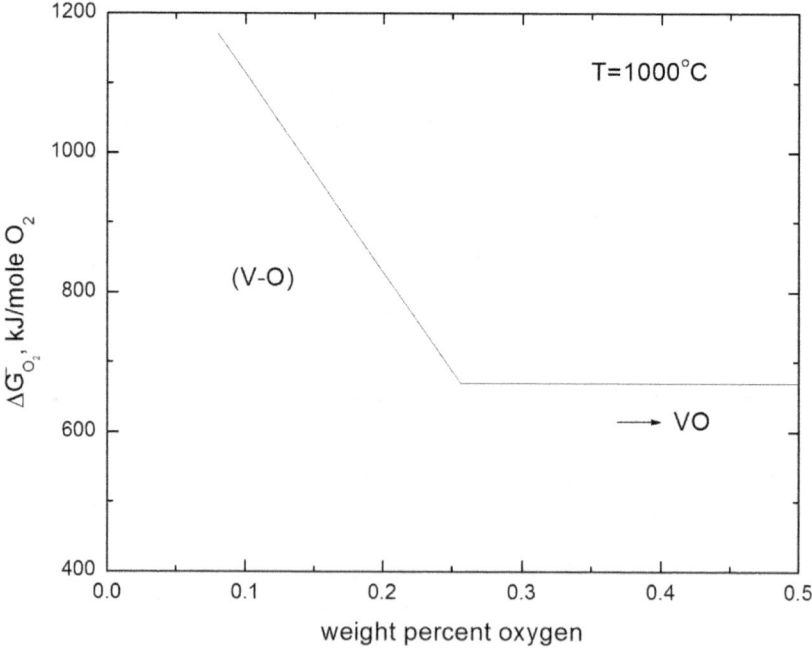

FIGURE 2.10 Partial molar free energy of oxygen in vanadium–oxygen system (From Allen et al. 1951).

the metal melt increases, it becomes more reactive. At 1000°C, the partial molar free energy of oxygen in vanadium–oxygen solid solution containing 0.1 wt% oxygen is comparable to the partial molar free energy of oxygen in titanium–oxygen solid solution containing 0.01 wt% oxygen (~1100 kJ/mole O_2). Contamination of oxygen from the oxide crucible is likely to be a more serious issue in vanadium than in titanium. The melting point of vanadium (1910°C) is also considerably higher than that of titanium (1668°C).

2.5 COMPONENTS OF INTERACTION

The overall metal–crucible interaction consists of the following main events (Fashu et al. 2020): (i) wetting of the crucible by the melt, (ii) penetration of the melt into the crucible to various depths, (iii) reaction occurring at the surfaces in contact, including the interior of pores, and (iv) products of reaction influencing the physical integrity of the crucible and properties of the penetrating melt. These events occur in a loosely sequenced manner and somewhat cyclically. It is the rate at which interaction occurs vis-à-vis the duration of the main process that is carried out in the crucible that finally determines whether the interaction can be lived with or should be remedied.

On melting, the metal's contact with the surface of the crucible is enhanced, and the liquid metal can also penetrate the pores. The extent of such contact is determined by the wetting between the liquid and the crucible material, fluidity of the melt, pore sizes and their interconnectivity, and the presence of other pathways for melt entry. Another factor is the hydrostatic head at the zone of contact between the liquid and the crucible. If the fluidity is high enough and if good wetting exists between the melt and the crucible, the melt can penetrate into the crucible by capillary forces, access and attack the grain boundaries (Figure 2.11) and weaken the bond between the grain and the crucible body. As noted in the general description, eventually the grain detaches from the crucible and is entrained into the melt. The

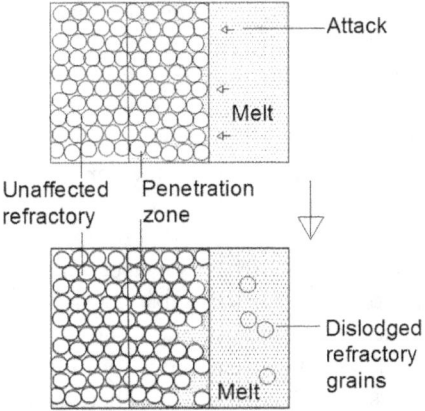

FIGURE 2.11 Melt penetration and physical erosion (schematic).

process is repeated on the next grain, and so on. Once in the melt, the particles follow the motion of the melt and are carried away from their initial location.

The removal of the particles results in the enlargement of the channel, giving access to the liquid to enter into the crucible surface region and eventually, more and more into the bulk. One thing leads to the next, and the degradation of the crucible by the combination of physical and chemical actions continues. The process is perpetuated by the continuous increase in the conductance of the channels in the crucible surface, the consequent continuous increase in contact area between the crucible and the liquid, and easier transport of entrained particles into the bulk of the melt. The eroded ceramic particles may either survive in the melt or be dissolved in it.

Generally, the interaction between a ceramic crucible and molten metal involves chemical reactions on the crucible wall and physical erosion. In such a situation, erosion will complement chemical attack by removing the reaction products from the surface, in the process exposing fresh surface, and the process goes on. The events are somewhat similar to the familiar case of atmospheric oxidation, wherein the oxide layer is non-protective and falls off on its own. The physical erosion in the present case is an event brought about by the conditions prevailing in the melt, such as free or forced convection, liquid currents generated by inductive stirring accompanying the heating process, or currents of liquid and bubbles generated by degassing processes in the melt or even by the stirring action of mechanical stirrers.

A simple possibility is that the melt enters the pore till it reaches the freeze line and solidifies. No reaction occurs, and as the pores are filled, further penetration of the melt also stops. The consequence is more thermal than chemical. The thermal conductivity of the infiltrated portion often increases, and this will affect the thermal balance of the system. An additional problem will arise when the crucible is cooled and the infiltrated melt solidifies completely. Depending on the thermal expansion characteristics of the solidified melt vis-à-vis the original crucible material, the integrity of the crucible will be affected.

After permeating the pores, if the melt interacts with the crucible material, many possibilities emerge. The sequence is different when reaction follows penetration into the pores of the crucible wall. The reaction weakens the crucible structure when one or more of its constituents, usually the binder or matrix, is dissolved in the melt or otherwise transformed. The binding becomes weaker, and particles become loose and are released into the melt. The pore size increases, and the process escalates. The reverse may happen if the dissolution of crucible material into the melt increases the melt viscosity. The progression of the melt into the pores slows down, and eventually, the permeation becomes sluggish or stops.

When the interaction between the melt and pore walls results in a new chemical product, and the physical properties of the products are different, things begin to happen. The property to look for first is the density or specific volume of the product. If the density of the product is higher than that of the original crucible material, the product occupies less volume than the material it consumed; the pore size increases, and new pore-like pathways may also be created. There will be escalation of penetration. If the density of the product is lower than that of the original crucible material, the product occupies more volume than the material it consumed. One beneficial thing that may happen is that if the density difference is just right, the formation of the product will cause closure or packing of the pores due to the additional space needed. Continued infiltration will stop. On the other hand, if the volume expansion is greater than can be accommodated in the pores, the product will end up disintegrating the crucible material in the infiltrated zone. Additional damage may be caused by the product's thermal expansion and thermal conductivity characteristics.

When the characteristics of the product of initial interaction are right, in the best case, the product forms as a compatible continuous and dense layer on the crucible material and effectively protects it from any further attack, just like the protective oxide film that forms on certain metallic materials during oxidation.

It all begins with the melt contacting the crucible surface and proceeds with a simple dissolution of one or more of its components or a chemical reaction with the crucible material followed by dissolution. The beginning is always a contact, and a close contact between the melt and the crucible is determined by how well the liquid wets the solid.

2.6   WETTING AND CONTACT ANGLE

Wettability is the ability of a liquid to spread over a solid surface. As it describes the extent of intimate contact between a liquid and a solid, it is an important characteristic for metal–crucible interactions. Generally, greater wettability results when the adhesive force between the liquid and the solid is more than the cohesive force (surface tension) within the liquid.

Figure 2.12 shows two-dimensional cross sections of a droplet on a solid surface. The left most droplet does not spread over the solid surface and is said to have a large contact angle. The right most droplet has spread well and has a low contact angle. The characteristics of the middle drop are in between. This spreading is known as 'wetting', and a droplet either 'wets' or 'dewets' when in contact with a solid surface.

The angle between the droplet outline and the solid surface is the contact angle. The angle is taken through the droplet to the tangent drawn to the droplet

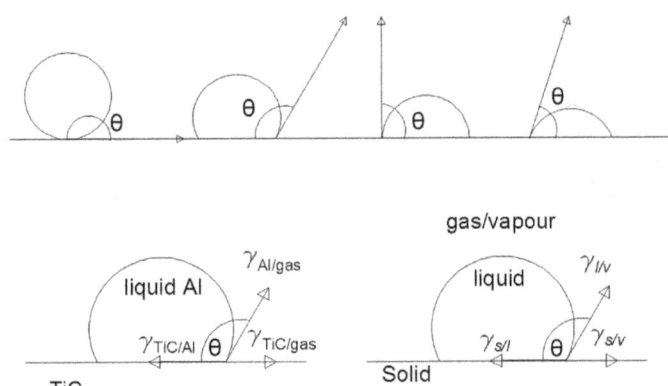

FIGURE 2.12   Wetting and contact angle.

outline at the point where the droplet periphery contacts the solid surface. A contact angle gives an indication of how well (or how poorly) a liquid will spread over a surface. A contact angle can be large or small, depending on the physical properties of the materials involved. An acute contact angle represents wetting, and an obtuse one represents dewetting. The smallest acute contact angle indicates maximum wetting.

There are three boundaries to consider when a droplet is in contact with a solid surface: the solid, the liquid, and the vapour (usually air) surrounding them. Figure 2.12 shows a force diagram of the point at which a droplet edge meets a solid surface. The three arrows represent the forces exerted by the surface tensions at three interfaces: liquid–surface(solid), liquid–vapour and solid–vapour (shown as $sl=$TiC/Al, $lv=$Al/gas, $sv=$TiC/gas,).

Each force is pulling away from the equilibrium point, so if the droplet is in equilibrium, then the forces are balanced and can be described by the following equation:

$$\gamma_{sv} = \gamma_{sl} + \gamma_{lv}\cos\theta \qquad (2.9)$$

Where $\cos\theta$ gives the x-component of the liquid–vapour surface tension. This can be re-arranged to give:

$$\cos\theta = \left(\gamma_{sv} - \gamma_{sl}\right)/\gamma_{lv} \qquad (2.10)$$

If $\gamma_{sv} < \gamma_{sl}$, then $\cos\theta$ will be negative, and θ is therefore $>90°$ and the droplet dewets. This can occur with a low-surface-energy solid or a high-surface-tension liquid (such as water).

If $\gamma_{sv} > \gamma_{ls}$, $\cos\theta$ is positive, and θ is $<90°$ and the droplet wets. This can occur with a high-surface-energy solid (such as a metal) or a low-surface-tension liquid.

(If γ_{sl} is larger than γ_{sv}, the contact angle θ is larger than $90°$; if the reverse is true, θ is acute. If $\gamma_{sv} = \gamma_{sl} + \gamma_{lv}$, the liquid will spread out over the solid surface.)

$$\cos\theta = \left(\gamma_{sv} - \gamma_{sl}\right)/\gamma_{lv} \qquad (2.11)$$

$$\gamma_{sl} = \gamma_{sv} - \gamma_{lv}\cos\theta \qquad (2.12)$$

The interface energy can be determined from the contact angle and liquid surface tension if the solid surface energy is known. The work of adhesion represents the decrease in energy in bringing together a unit area of liquid surface and a unit area of solid surface to form a unit area of interface. The work of adhesion can also be calculated directly from experimental data:

$$W_{ad} = (\gamma_{lv} + \gamma_{sv}) - \gamma_{sl} = \gamma_{lv}(1+\cos\theta) \qquad (2.13)$$

The work of adhesion is maximum when the wetting is perfect. In other words, there is maximum decrease in energy when a unit area of liquid surface and a unit area of solid surface are brought together under conditions of perfect wetting (contact angle = 0) to form a unit area of interface.

The wettability of a given solid/liquid couple can be assessed by measuring the contact angle in a system consisting of a drop resting on a flat solid surface, in a given atmosphere, by the sessile drop method. A general (feature) limitation of the information thus obtained has been highlighted by Lopez and Kennedy (2006).

The wettability of TiC by molten aluminium has been investigated in various temperature ranges by many groups. More often than not, the measured contact angle changes with time even while experimental conditions are maintained the same. The actual interface chemistry and surface-active impurities originally present in the solid and melt phases tend to play a decisive role (Fashu et al. 2020). The contact angle at any point of time is determined by the real-time composition of the contacting surfaces. This is also highlighted in the results obtained by Durov et al. (2005).

Wetting of zirconia by metals was studied by Durov et al. (2005). Single crystals of 3% yttria-stabilized zirconia and some polycrystalline ZrO_2 ceramic samples were the substrates. Reactive metals (aluminium and vanadium) wet zirconia; less reactive metals (tin, copper, silver) do not wet it. But, Durov et al. (2005) noted that relating wetting to affinity for oxygen is not straightforward; silicon does not wet the zirconia, whereas germanium wets it. The interaction of germanium with zirconium plays a role, because stable intermetallics exist in the Ge–Zr system. A similar situation exists for the interaction between germanium and rare earth oxides.

The wetting of zirconia by the alloys Cu–Ga and Cu–Ge and also ternaries with titanium, zirconium and vanadium was studied by Durov et al. (2005). The addition of titanium and zirconium improves the wetting. Vanadium, even at low concentrations, steeply decreases the contact angle, and the effect is greater than that due to Ti and Zr.

When the wetting of different types of zirconia ceramic by a metal melt is compared, monocrystalline zirconia is wetted more (smaller contact angle) than polycrystalline zirconia. The structure of zirconia material has a larger influence on the wetting than its composition.

Black zirconia (non-stoichiometric oxide) is wetted by melts of noble metals and alloys (Cu, Sn, Ag, Au, Pd, Pt, Cu–Ga, Pd–Rh) somewhat better (contact angles are smaller) than white zirconia (contact angles are larger). Remarkably, after contact with gold, platinum and palladium, black samples regained their white colour. On examination of the distribution of elements on ZrO_2/Pd and ZrO_2/Pt interfaces, it was revealed that the excess zirconium in oxygen-deficit zirconia passes into the noble metal melt, resulting in removal of the oxygen deficit and reversion of the colour to white, which is characteristic of stoichiometric zirconia. While noble metals restore the stoichiometry of non-stoichiometric zirconia (turning black to white), active metals make the stoichiometric zirconia non-stoichiometric (turning white to black).

Wetting, in turn, determines the ease with which the melt will penetrate the pores (invariably present) in the crucible and set the stage for further interaction. Non-wettability is a property much desired in crucibles. When the wettability is poor between the ceramic crucible and the melt, there will be a decrease in infiltration and eventual degradation of the refractory by the melt. To run a preliminary check on crucible–melt interaction severity and wettability, sessile drop tests may be performed on potential crucible materials.

The wetting of graphite/alumina substrates by molten iron at 1600°C was studied by Zhao and Sahajwalla (2003). Pure graphite substrate was wetted by liquid iron with the contact angle varying from 64° at the beginning to 38° at the end. The presence of alumina had a strong influence on contact angle for carbon–alumina substrates. The contact angle showed a sharp change from wetting to almost non-wetting as the alumina content was increased from about 17% to 23%. Alumina also retarded carbon dissolution. The influence of alumina content on the contact angle between liquid iron and alumina–carbon mixtures is summarized in Figure 2.13. Mixtures of alumina and petroleum coke in liquid iron displayed non-wetting behaviour. Orsten and Oeters (1986) also reported slower dissolution of carbon in less ordered forms of carbon material (like petroleum coke) as compared with ordered counterparts like graphite.

That the contact angle is so dependent on the real-time chemistry of the interface is both good news and bad news in the design of crucible compositions. It is not good news because the impurities, especially surface-active ones, in crucible compositions can end up behaving in unpredictable ways with expensive consequences. It is good news because the window is open for deliberate choice of impurities (now called additives) to modify the contact angle by analogy and experimentation.

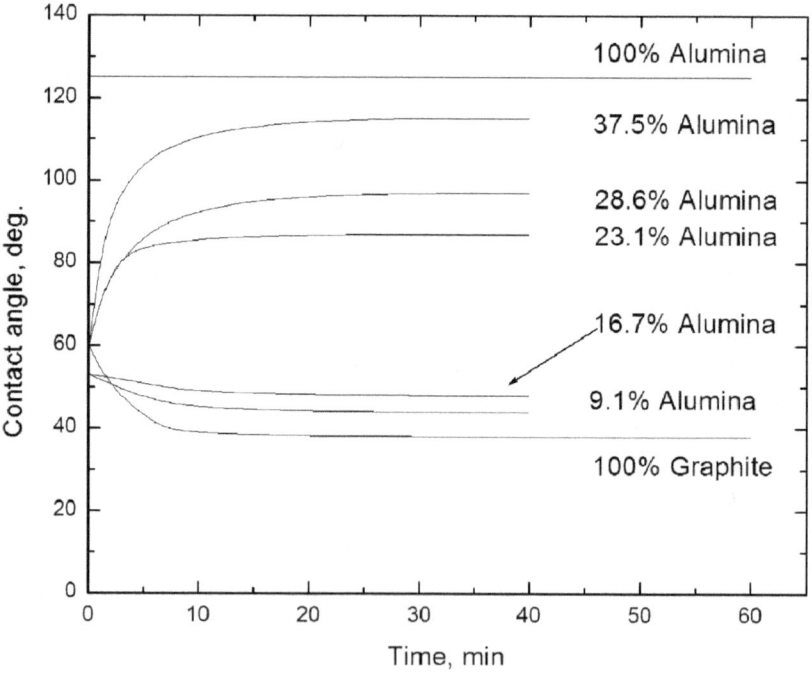

FIGURE 2.13 Contact angle of liquid iron with alumina–graphite (From Zhao, L. and Sahajwalla, V. 2003).

2.7 POROSITY

Almost all refractory materials are porous. The pores may occupy 1–80% of the material. The pores may be open (connected with the atmosphere) or closed (isolated from the atmosphere and each other). Refractory materials always contain some number of closed pores.

The ratio of open to closed porosity in refractory ceramics is, to some extent, determined by the sintering process. In bricks with porosity exceeding 15%, the proportion of closed pores is normally small (<ca. 3%). The pores of the ceramic body, fabricated from non-porous starting grains, remain open until the relative density during sintering reaches 83–85%, after which closed pores will begin to appear. At a relative density of 95%, almost all the pores are closed.

Generally, the porosity of traditional fireclay refractories is 22–25%, in modern aluminosilicates it is 13–15% and in silicon carbide refractories it is usually 12–18%. The porosity of low-cement and ultralow-cement castables is 12–15%, but the proportion of permeable pores is lower. Cathode carbon refractories, used in Hall–Héroult cells, have porosities of 15–22%, and almost all pores are open and permeable. In carbon materials, for example, almost all the pores are open (Yurkov 2017).

Different refractory materials may have approximately equal values of porosity but a different ratio of permeable pores, a different pore size distribution and a different pore structure.

Porosity and density are important and to some extent controllable characteristics of refractory materials. The mechanical, thermal and other properties of refractories depend on them.

Porosity influences the strength properties of refractories. A certain porosity is usually necessary for the refractory to have the adequate thermal shock resistance, while corrosion resistance or resistance to metal melt or molten slag attack becomes better with a decrease in porosity. Both these conflicting requirements are to be met simultaneously. Difficult, but it has to be done in many instances.

2.8 CAPILLARY PENETRATION

The primary condition for metal–crucible interaction is the close contact between the melt and the crucible over a large interface area. Apart from a small contact angle, penetration of the liquid metal into a porous crucible structure is a major factor. Porosity, pore size, pore size distribution, fluidity of the melt and the hydrostatic head at the zone of contact between the liquid and the crucible are the key factors.

The earliest systematic theoretical description of liquid flow in cylindrical capillaries, taking into account capillary pressure, gravity and outside pressure, is attributed to Washburn (1921). A very elegant description of the whole process was given by Kaptay et al. (2004).

The presumption in the classical capillary theory of penetration is that the refractory consists of a large number of vertical, cylindrical capillaries of identical internal radii r (Figure 2.14). There are three pressure vectors acting on the liquid column in those capillaries:

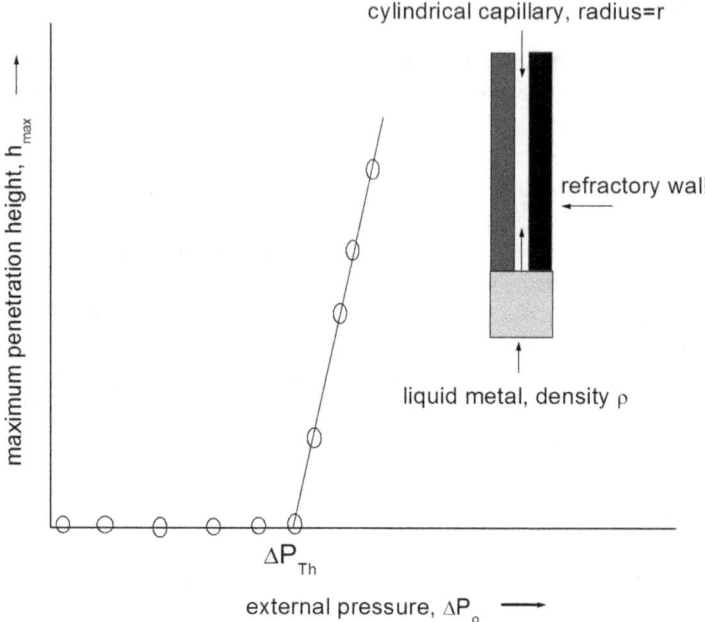

FIGURE 2.14   The classical capillary penetration model (From Kaptay, G., Matsushita, T., Mukai, K., and Ohuchi, T. 2004).

1. If the liquid wets the inner wall of the capillary, contact angle, θ, is less than 90, the capillary pressure (Pc), pushes the liquid up (capillary sucks in the liquid). When the contact angle is more than 90, the capillary pressure pushes the liquid out of the capillary

$$P_C = (2\gamma \cos \theta)/r \tag{2.14}$$

   where γ is the surface tension of the liquid (J/m²).
2. The pressure due to gravity (P_g), always acts to push the liquid out of the capillary:

$$P_g = -\rho gh \tag{2.15}$$

where
   ρ = density of the liquid metal (kg/m³);
   g = 9.81 m/s², the acceleration due to gravity; and
   h = the penetration height, measured from the bottom of the refractory (m).

In experiments, it is usual to provide the outside pressure difference (ΔP_0) so that the liquid rises into the capillary. The liquid column in the capillary eventually comes to equilibrium if the sum of these three pressures (P_S) is zero

$$P_S = P_C + P_g + \Delta P_0 = 0 \tag{2.16}$$

When liquid column height is less than 1 mm, P_g is negligible compared with P_C, and the condition for penetration to begin becomes

$$P_C + \Delta P_0 \geq 0 \tag{2.17}$$

Substituting Eq. (2.14) into Eq. (2.17),

$$\Delta P_0 + (2\gamma \cos\theta)/r \geq 0 \tag{2.18}$$

From condition (2.18), two results follow.

1. Spontaneous penetration occurs (i.e. at $\Delta Po = 0$) if the contact angle is acute. This contact angle, known as the threshold contact angle of penetration, is usually different from 90°. However, for capillaries with parallel vertical walls of any cross section, it is always 90°

2. When the contact angle is larger than the threshold contact angle of penetration, for any penetration to occur, the outside pressure needs to be higher than the threshold pressure. The threshold pressure can be expressed as

$$\Delta P_{th} = -(2\gamma \cos\theta)/r \tag{2.19}$$

When $\Delta P_o > \Delta P_{th}$, penetration will start. As the liquid column height becomes significant, P_g can no longer be neglected compared with P_c. Then, the following condition of equilibrium can be obtained by substituting Eqs. (2.14) and (2.15) into Eq. (2.16):

$$(2\gamma \cos\theta)/r + \Delta P_0 - \rho g h = 0 \tag{2.20}$$

From Eq. (2.20), the equilibrium (or maximum) penetration height (h_{max}) can be expressed as

$$h_{max} = \Delta P_0 / \rho g + 2\gamma \cos\theta / r\rho g = (\Delta P_0 - \Delta P_{Th})/\rho g \tag{2.21}$$

Zhang and Lee (1999, 2004) estimated the depth of slag penetration using the equation:

$$L^2 = (r\cos\theta)\gamma t / 2\mu \tag{2.22}$$

where L = penetration depth, r = radius of the pores, θ = contact angle, γ = surface tension of liquid, t = interaction time and μ = viscosity of liquid slag.

It is apparent that a large pore size, wettable crucible surface, high melt surface tension, less viscous melt and long interaction times cause deeper melt penetration into the crucible. A decrease in pore radius and increase in viscosity of the melt can effectively stop any penetration. These two are useful handles in liquid ingestion control.

   The slag viscosity is a function of both temperature and chemical composition of slag. The penetration depth can be decreased by using a refractory material of low thermal conductivity, which creates a thermal gradient between the hot face and the cold face of the lining. As the slag penetrates into the refractory, the viscosity increases with depth of penetration (because the temperature decreases). After some distance, the temperature drop and consequent viscosity increase stop further penetration. The other factor to reduce the viscosity is the change of chemistry of the slag during operation. The spinel structure, for example, is able to absorb Fe and Mn ions from the slag. As a result, the slag becomes rich in SiO_2, and viscosity increases. Any other reaction that results in an increase in the concentration of silica in the slag increases its viscosity. An increase in the concentration of FeO, MnO makes the slag more fluid.

   When discussing pore size dimensions in a refractory, it is necessary to mention the surface tension of the melt. The penetration of the melt in the refractory depends not only on the pore radii and viscosity but also on the surface tension of the melt, which is in any case also a factor that determines the contact angle of wetting. The typical threshold pore diameter above which penetration of various metal melts becomes serious is given by Rouchka (2001) and is shown in Figure 2.15.

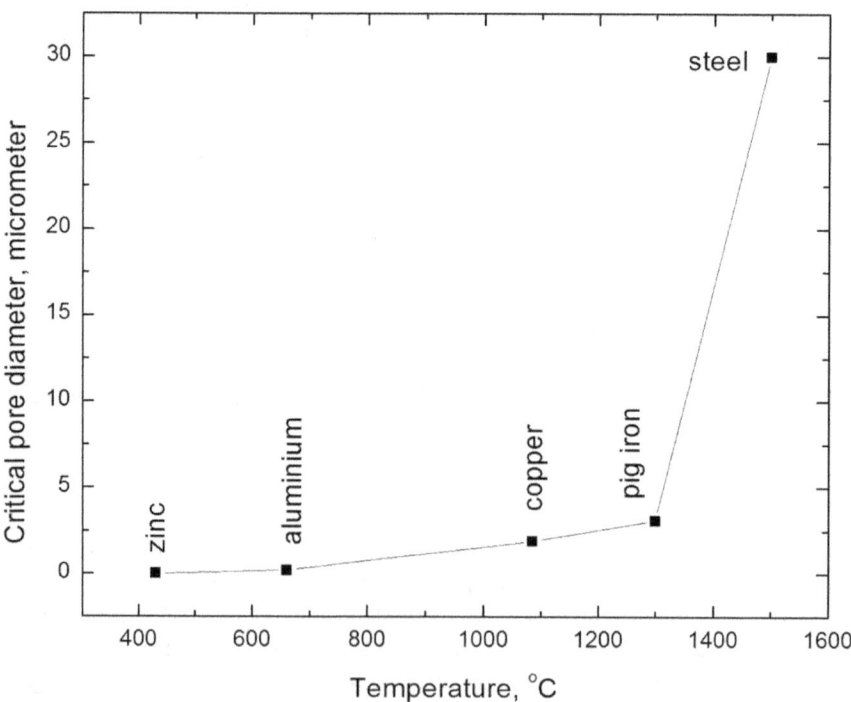

FIGURE 2.15   Critical pore diameter for infiltration at the melting point of metals (From Routschka, G. (Ed.). 2001).

2.9   PHYSICAL PROPERTIES OF REFRACTORIES

Apart from chemical behaviour, certain physical properties of both the refractories and the melts are of key importance in determining the extent and rate of metal–crucible interactions. Foremost among these are the melting point of refractories and the freezing point of melts. Then, there are properties like thermal expansion coefficient, thermal conductivity, high temperature strength, thermal shock resistance, polymorphism and many more of contextual relevance. Data on melting, softening points, shock resistance and Ellingham free energy of formation for selected potential refractory materials are listed in Table 2.2.

2.10   THERMAL EXPANSION

Like all solids, refractories expand on heating. Figure 2.16 shows the expansion behaviour of some refractory types (Fruehan 1998). The shape of the expansion curve and the magnitude of the expansion are distinctly different for silica. The complex mineral makeup of silica brick confers on it a uniquely different expansion behaviour. Other refractories show a constant linear behaviour, except that the slope differs appreciably among the materials. In any refractory assembly, allowance is always made for accommodating such expansions in service (Fruehan 1998).

Thermal expansion curves for magnesite–carbon bricks with various antioxidant additives and plain magnesite are shown in Figure 2.17. Their shapes are more or less similar. In the temperature range 370–540°C, the formation of a glassy carbon

TABLE 2.2
Properties of Refractory Ceramics for High-Temperature Applications

Ceramics	ΔG^0_{2000} kJ/mole O_2	Thermal conductivity (W/m/K)	Coefficient of thermal expansion ($\times 10^{-6}$/K), 20–1000°C	Melting (softening) temperature, °C	Thermal shock resistance factor
SiO_2	−550.7	1.2–1.4	0.5–0.8	1710 (1280)	93
Al_2O_3	−702.6	38	8.0	2072 (1540)	19–26
MgO	−660.0	30–60	9–12	2852 (2100)	18
CaO	−809.0	30	15.2	2572 (1950)	–
ZrO_2	−725.5	2.5	10.0	2715 (2010)	12–13
Y_2O_3	−891.3	8–12	8.1	2425 (1855)	–
AlN	−185.6*	280	4.5	2200	37
BeO	−800.0	370	8	2507	–
BN	−159.1[a]	1300	7.5	2200	–
Graphite	–	140	0.6–5.2	3530 (2680)	–

Source: Fashu et al., 2020.

[a] kJ/mole N_2

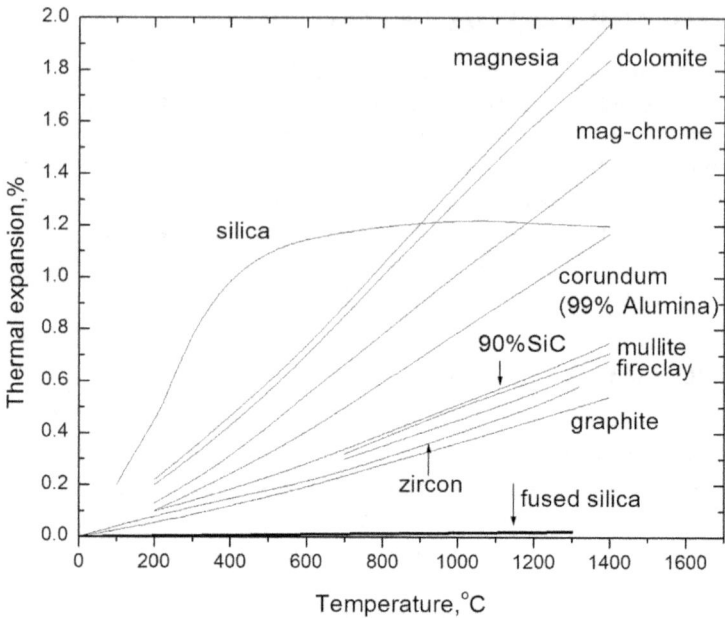

FIGURE 2.16 Thermal expansion behaviour of refractories (From Fruehan, R.J. 1988, Routschka, G. 2001).

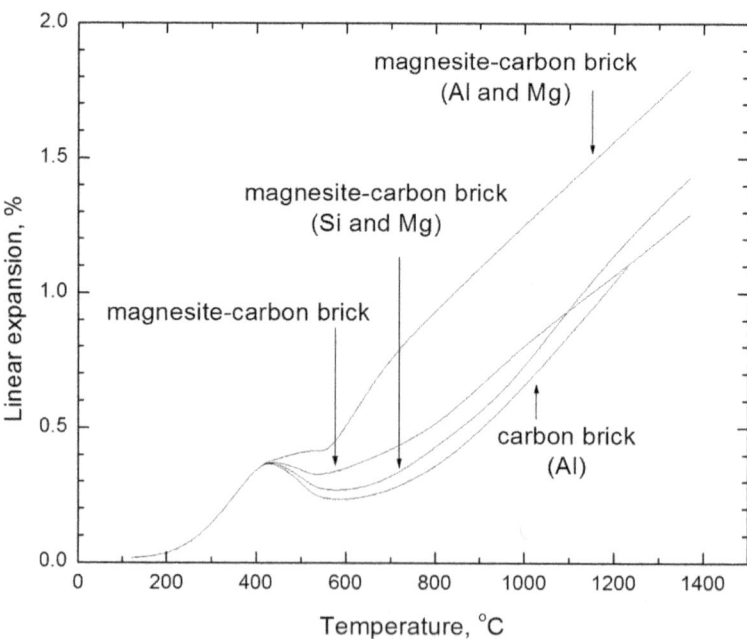

FIGURE 2.17 Thermal expansion behaviour of magnesite–carbon bricks (From Fruehan, R.J. 1988, Routschka, G. 2001).

bond causes densification of the structure and shrinkage. In refractories that contain metals, thermal expansion at high temperatures increases significantly with increase in metal content.

Crystals with a cubic lattice (CaO, MgO) expand equally along all axes (isotropic). The thermal expansion coefficients for such materials are $6-15 \times 10^{-6}$ K^{-1}. The coefficients of thermal expansion are different along different axes for anisotropic crystals with low symmetry, but this difference becomes smaller with increase in temperature. Materials with strong chemical bonds (silicon carbide, titanium diboride, diamond) have low values of linear coefficients of thermal expansion. Pores do not affect the values of linear coefficients of thermal expansion if the continuous media are solid particles. Thermal expansion is a key component in determining thermal shock resistance.

2.11   THERMAL CONDUCTIVITY

As insulators, refractory materials have always been used to conserve heat for the charge and protect the metal shell of reactor vessels from overheating. The resistance to heat flow is the reason for their selection in many applications. Figures 2.18 and 2.19 show thermal conductivity curves for several refractories (Fruehan 1998).

The hot face refractories in a particular application must be able to withstand the higher temperatures that will result when layers of insulating backup materials are

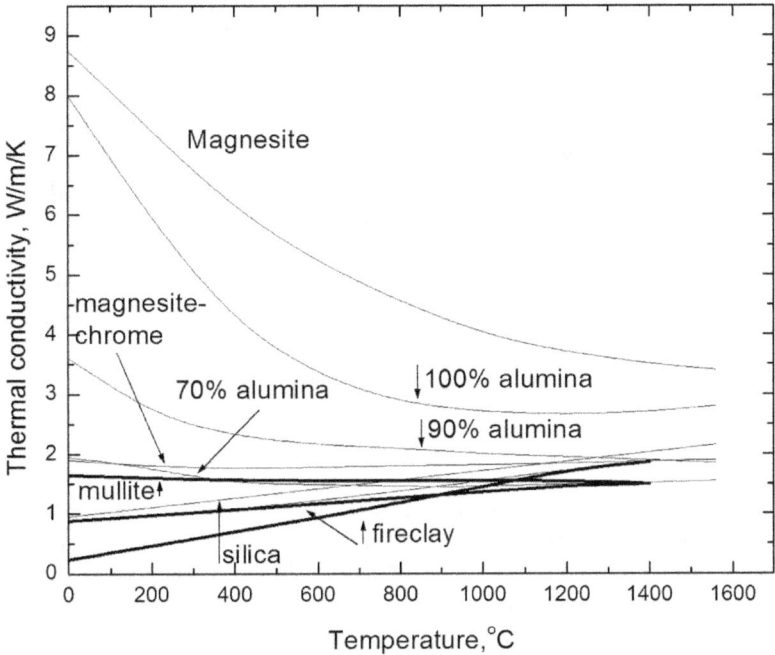

FIGURE 2.18   Thermal conductivity of various refractories I (From Fruehan, R.J. 1988, Routschka, G. 2001).

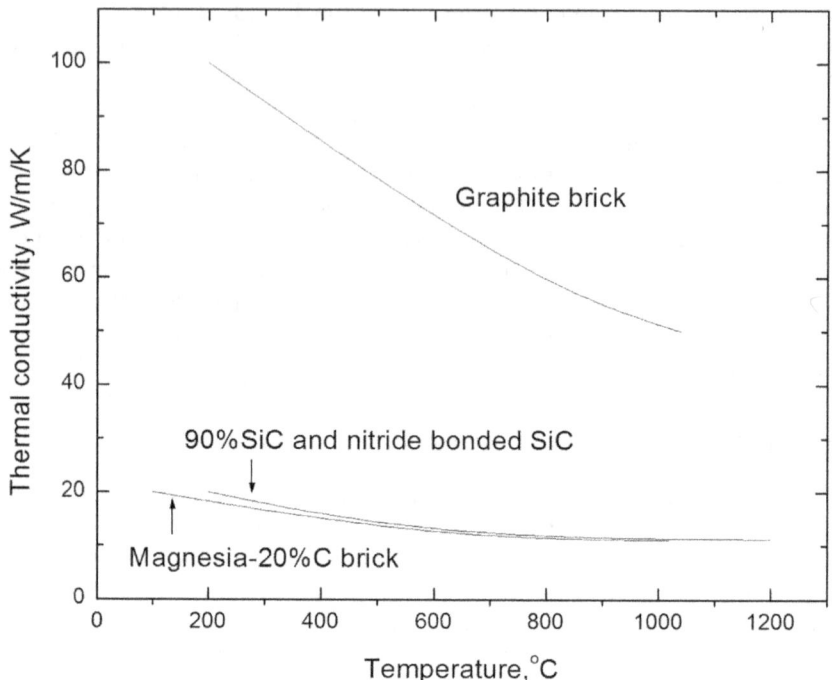

FIGURE 2.19   Thermal conductivity of various refractories II (From Fruehan, R.J. 1988, Routschka, G. 2001).

added. Insulation increases the depth of penetration and chemical attack on the hot face layer. Most insulating materials are prone to attack on direct exposure to metal or slag, vaporized process components (alkali, sulphur compounds, acids) or their condensates. In certain applications, high-conductivity refractories are used to facilitate cooling of the refractory lining and retard further refractory wear. For example, in a bosh construction, graphite and semi-graphite materials with conductivities of 70–80 and 30–35 $Wm^{-1}K^{-1}$, respectively, at 1000°C are used. The heat conducted is finally extracted through the copper coolers. Thermal conductivity, like thermal expansion, is also a key factor in determining thermal shock resistance.

2.12   THERMAL SHOCK

Thermal shock resistance is a complex issue. Thermal shock or spalling is caused by stresses that develop from uneven rates of expansion and contraction within the refractory, caused by rapid temperature changes or high inherent temperature gradients in a refractory. A qualitative prediction of the resistance of materials to fracture by thermal shock can be expressed by the factor (Fruehan 1998):

$$\frac{ks}{\alpha E}$$

Where k = thermal conductivity,
   s = tensile strength,
   α = coefficient of thermal expansion,
   E = modulus of elasticity.

According to this factor, the material should have a high thermal shock resistance if it has a low elastic modulus, a low value of the coefficient of linear thermal expansion, and high strength as well as high thermal conductivity. The higher the value of the factor in the expression, the higher the predicted thermal shock resistance of the material. A substantial proportion of all refractory failures are traceable to thermal shock issues. Refractories are damaged by poor thermal shock resistance at temperatures much lower than their regular service temperatures. Thermal stresses may be tensile, compressive or shear; due to thermal stresses, refractory materials may chip and split off the linings. Thermal shock resistance is not a physical property of a material; it depends on many physical characteristics of the material, but also on the shape size of the refractory part and the cooling and heating conditions.

2.13 CORROSION OF REFRACTORIES

Corrosive attack on the crucible refractories in a metal processing unit may be mounted by the molten metal, molten slag or the process gas. Often, it is the combination of all three, each contributing to the overall degradation. This is apart from, and in addition to, the thermal and mechanical factors that cause weakening or failure of refractories. The most significant and life-limiting attack is by the slag. Even the corrosion by the metal melt becomes more serious if it acquires characteristics of a slag, such as when it is oxidized or laden with oxides. The attack can happen in essentially two ways. The following description is fairly general and reflects the essential features of refractory degradation, be it happening in the lining of a large furnace or in a compact system comprising a crucible and melt in a controlled environment (Lee and Zhang 2004).

Direct (also known as congruent or homogeneous) attack is controlled by the reaction rate at the slag–refractory interface or the rate of transport of the reactant through the liquid film adjacent to the interface. Indirect (incongruent or heterogeneous) attack is controlled by diffusive transport of species to the reaction site through the slag or through a new solid phase, which forms at the original slag–refractory interface. This may lead to passive corrosion. The composition and hence, the viscosity of the local liquid slag adjacent to the solid refractory have a critical influence on corrosion. As is true with most of the rate processes, it is seldom that one process or other exclusively controls the rate. Usually, one predominates over the other. Penetration and corrosion can be controlled either through the local slag composition, via the refractory or the bulk slag, or by microstructural control of the refractory, e.g. by internal generation of dense layers or external deposition/generation of passive coatings. Selective corrosion may also occur, in which only certain phases in the solid are attacked. A good example of this is the decarburization of carbon-containing refractories, which occurs by several mechanisms, including dissolution of carbon into molten steel. Once the carbon has been removed, the refractory is open to be wetted by the slag, so that penetration and spalling of the decarburized layer can occur.

In direct, congruent or homogeneous dissolution, atoms from the solid dissolve directly into the liquid melt. Direct dissolution can be reaction or interface controlled when the diffusivity of reaction products is faster than the rate of chemical reaction at the interface. The dissolution process is then directly controlled by the reaction, whose initial rate can be expressed by

$$J = K\left(A_C / A_O\right)C_m \tag{2.23}$$

Where J is the dissolution rate (g/cm/s), K is the rate constant, A_C is the actual area of refractory (cm^2), A_O is the apparent area of refractory (cm^2), and C_m is the concentration of reactant in the melt (g/cm^3). Surface irregularities such as grooves and porosity, which increase the ratio of Ac to Ao in Eq. (2.23), become important. For this simple treatment, stirring of the melt has no apparent effect on dissolution rate. The concentration of the reactant at the interface (considered to be the same as the concentration in the bulk) and the area available for interaction are the only two things that matter, and these are equally important.

For direct dissolution to continue, the products diffuse away from the interface at a rate proportional to $t^{1/2}$ as reactants are depleted and dissolved species build up in the absence of liquid convection or stirring (Cooper and Kingery 1959). In a situation here the rate of removal of reaction products by diffusion is slower than the rate of chemical reaction, a solute-rich boundary layer builds up, whose interface with the refractory is saturated with reaction products. The diffusion of reactants to, or the products away from, the interface through the boundary layer then governs the dissolution process. If the boundary layer leads to formation of a solid interlayer, this is termed indirect, incongruent or heterogeneous dissolution, and the rate of corrosion can be expressed in terms of the Nernst equation:

$$J = D(C_S / C_m) / \delta \tag{2.24}$$

Where D is the diffusion coefficient (in cm^2/s), Cs is the saturation concentration of refractory component in the melt (g/cm^3), and δ is the effective boundary layer thickness (cm), which is defined as

$$\delta = \left(C_S - C_m\right) / (\delta c / \delta x) \tag{2.25}$$

Where dc/dx is the concentration gradient over the interface. For diffusion-controlled direct dissolution, diffusion of the product through the melt boundary layer is considered. However, once saturation occurs, and a solid interlayer has precipitated between the melt and the refractory, this is the diffusion-controlled indirect process, and solid-state diffusion of the reaction species from the melt through the solid interlayer must be considered.

Stirring the melt or rotating the refractory sample enhances the rate of indirect dissolution (or effectively converts it to direct dissolution) by reducing the thickness of any liquid boundary layer (or breaking up any solid layer). Apart from dynamic viscosity, the value of $(C_s - C_m)$ has a great influence on the dissolution rate. If the refractory oxide in the bulk slag has been saturated, then $J = 0$. Naturally, to minimize the dissolution rate, $(C_s - C_m)$ must be minimized. For example, increased MgO content in slag decreases the corrosion of periclase, the primary phase in many basic refractories. If $C_m = 0$, then $(C_s - C_m)$ reaches a maximum value, and so does the dissolution rate. Consequently, the solubility of refractory oxides in molten slag and the saturation concentration at the interface between refractory and slag are very important in corrosion of refractories.

Slag viscosity has significant effects on both the slag penetration and the refractory dissolution. A more fluid slag will be more penetrating and more likely to dissolve the solid refractory. Furthermore, if dissolving the refractory in the liquid leads to increased viscosity, then mass transport through the melt layer next to it will be slower, so that the melt layer becomes progressively saturated, giving rise to diffusion control, i.e. indirect dissolution. On the other hand, if the viscosity of the melt layer is decreased, then diffusion through it becomes more rapid, no saturated layer forms, and reaction control or direct dissolution occurs. Raising the temperature to give a more fluid slag can then lead to a change from indirect to direct dissolution.

Even in the ancient process of cupellation, the role of melt viscosity for process success was highlighted. Basically, a melt containing lead oxide and silver at 960 to 1000°C had to be separated physically. Liquid lead oxide was to be absorbed into a special porous crucible, leaving behind silver. The crucible had to be made of inert and porous material rich in calcium or magnesium, such as shells, lime or bone ash. Only a calcareous lining would work, because lead reacts with silica (clay) to form lead silicate, and lead silicate is viscous and impedes the absorption of litharge. Lead does not react with the calcareous materials, and fluid lead oxide melt is readily captured by the crucible purely by the physical phenomena of absorption. Generally, a silicate enrichment in an oxide slag makes it viscous, and so does FeO depletion. This second aspect was in a way responsible for the delayed implementation of ironmaking with separate molten iron and slag tapping in shaft furnaces in the early days.

2.14 CHEMICAL DISSOLUTION IN CASTING

The principal difference between metal–crucible interaction during melting and during casting is that in the latter case, the duration of contact between the melt and crucible/mould is very limited, and the melt is relatively quiescent before it sets into a solid. Immediately after its formation (i.e. melting in the crucible or teeming into the mould), the melt tends to react with the crucible and dissolve the constituents of the crucible. For example, for an Y_2O_3 crucible, the dissolution reaction is

$$Y_2O_3 = 2Y_L + 3O_L \qquad (2.26)$$

Where L stands for melt, such as molten titanium or titanium alloy.

In conditions conducive to mass transport, the dissolution will be at the maximum rate, and the melt will soon attain a uniform concentration. The dimensions of the crucible are also a factor. When the sample is allowed to solidify inside the crucible, the interaction between the crucible and the melt develops somewhat differently. With the progress of solidification of the alloy, in regions that have become solid, mass transport can occur only by diffusion, and in portions that are still liquid, mixing occurs but to a much lower degree, because stirring and similar mass transport aids no longer operate. As a result, composition gradients develop between the surface and the bulk and can be discerned in the solidified material.

2.15 CORROSION TESTING OF REFRACTORIES

There are various tests for single refractory properties. Because most refractories wear by a complex combination of mechanisms, many simulated tests have been developed to study refractory behaviour for specific applications (Fruehan 1998; Lee and Zhang 2004). Tests have been used to create the environment a refractory is exposed to during service. They are categorized as dynamic, where the fluid moves relative to the refractory, and static, where there is no such motion.

In the button or sessile drop test, shaped slag is placed on a refractory substrate, heated to temperature and held for a fixed time to allow the slag to wet and react with the refractory. The technique is used to measure interface and surface energies in systems where the fluid is liquid and does not react with the solid. In the dipping, immersion or finger test, one or more cylindrical or square-pillar-shaped refractory samples are held in the corrosive slag for a certain period in an electric or induction furnace. In this method, the atmosphere is easily controlled, and by using a large volume of slag relative to the size of the samples, the composition variation of the slag, associated with rapid saturation with reaction products, can be minimized. These tests are isothermal, i.e. at the same temperature throughout their volume.

In the crucible, cavity, cup or brick test, a cored-out refractory brick is filled with slag and exposed to high temperature to promote slag–refractory interaction. In this simple method, many samples can be tested in a short time. However, it has limitations associated with static tests, i.e. no temperature gradient, rapid saturation of slag composition with reaction products (often, all slag is absorbed into the brick) and no slag flow. In tests conducted in an *induction* furnace, a temperature gradient can be established, and atmosphere and temperature are easily controlled. Rapid and vigorous corrosion can occur at the melt/slag line, since the less dense slag floats on the metal. An induction furnace test induces some motion (electromagnetic stirring) upon heating and so has some dynamic aspects. The results of such a test can better simulate the situation in a real furnace, except that flow is uncontrolled. Relative motion of liquid slag and solid refractory is particularly important in systems where indirect dissolution occurs. The motion can remove the boundary layer and enhance any active corrosion process. Dynamic tests are needed to simulate such behaviour. In fact, this can be a signature test of whether dissolution is indirect (in which case corrosion rate increases with motion) or direct (in which case motion has no effect).

The dipping test can be made dynamic by spinning the samples in the molten slag, a rotating finger test. A popular corrosion test is the rotary slag test, in which the slag is melted with a flame in a cylindrical refractory-lined drum rotating about a horizontal axis. The slag is refreshed periodically during the test by tilting to remove old slag, returning to level and adding new slag. In this method, many samples can be compared in a single test; a temperature gradient can also be established, and the composition and fluidity of the slag are partially controlled.

The chemical composition, hardness and microstructure at the metal–crucible interface can be measured by secondary ion mass spectrometry, scanning electron microscopy/energy-dispersive spectroscopy (SEM/EDS) and X-ray spectroscopy (XRS). Chemical composition measurement can be performed by quantitative EDS analysis with appropriate attachments using a scanning electron microscope. Overall oxygen content may be measured using the inert gas fusion (IGF) technique, and oxygen content variation from the surface to the inside of samples by secondary ion mass spectroscopy (SIMS).

In SIMS, a 4 keV 200 nA beam of Ar^+, for example, is focused in a 50 μm diameter area at a time, and secondary ions produced by sputtering are mass analysed by the attached quadrupole-type mass spectrometer. XRD identifies compounds present on the microstructure. Microhardness may also be evaluated with a hardness tester. Hardness is a popular indicator of material purity.

Used and or failed refractories are examined by chemical analyses and by microscopic methods for clues. X-ray, SEM, or electron probe techniques are used as part of postmortem examinations, and valuable insights are often gleaned, giving direction to effective corrective actions (Fruehan 1998).

2.16 SUMMARY

Metal–crucible interaction operates by chemical interaction between the material of the crucible and the melt, augmented by concurrent physical, mechanical and thermal processes. Chemical interaction, in turn, generates products that because of their own properties, may cause changes that also result in degradation. The key process, therefore, is the chemical reaction, and chemical reactions are governed by thermodynamics.

Ellingham diagrams, essentially the plot of standard free energy of formation of compounds as a function of temperature, do indicate whether a reaction may be expected between a compound of one metal and another metal in its elemental form, in the present case between the melt and the crucible material. In numerous cases, the predictions have been true and complete, and processes have been founded on that basis. In many other cases, the predictions do not reflect the real situation completely, because the data on the basis of which the diagrams have been prepared are in themselves incomplete. This is especially true for reactive metals that form extensive and stable solutions, and also oxides that exhibit non-stoichiometry. It is therefore necessary to consider the relative stabilities of the oxides, including the known hypostoichiometric oxides, and also that of the solutions the metal forms. Proceeding thus, as a test case, it has been highlighted that in contact with the most stable oxide crucible, yttria, near the melting point of titanium, the residual oxygen content in titanium cannot decrease below 0.05 wt.% oxygen. In contrast, a casual consideration of Ellingham plots may point to the possibility of titanium metal (by implication 'pure') being obtainable by reduction with many common reducing agents (metals, of course). By extension, this also means the possibility of melting titanium without contamination in many crucible materials. Experience with melting procedures has shown that this assertion does not reflect reality.

Even though the statement that at sufficiently high temperatures, everything reacts with everything else cannot be disputed, it can be tempered with many ifs and buts. These conditions, which are extremely important in real-life situations, are physical, structural, mechanical and thermal. A crucible material should possess, in addition to high thermodynamic (chemical) stability, high melting point, good strength, appropriate thermal expansion and thermal conductivity behaviour, and thus, thermal shock resistance. Some of these properties are interrelated. Tailoring of some of these properties by microstructure and porosity control has been resorted to more or less routinely in order to ensure that the chemical stability of the crucible material is fully utilized in actual application.

To interact, the melt and crucible have to be in close contact. This aspect is decided by the wetting characteristics of the melt–crucible combination. The contact angle is a measure of wetting and is determined by the interfacial tensions of the pairs: liquid–solid, liquid–vapour and solid–vapour. The contact angle or wettability is ultimately determined by the real-time chemical composition of the melt–solid interface. Impurities in the melt or crucible play a role, and this is the basis of anti-wetting agents. Anti-wetting additives act by controlling the interface chemistry and raising the contact angle. They have been used extensively with aluminium melts because of the temperatures involved. Nearly all anti-wetting additives decompose at high temperatures, and then the protection is lost (usually beyond 1000 to 1200°C). For this reason, porosity control may turn out to be the more reliable handle to obviate melt intrusion. However, modifying porosity may have other unintended consequences.

Corrosive attack on the refractory of the crucible is often due to the combined action of molten metal, slag and also process gas. The most significant and life-limiting attack is by the slag and also by metal that acquires the characteristics of slag. The attack proceeds in two ways. Direct dissolution is controlled by the rate of chemical reaction at the slag–refractory interface. Indirect attack is controlled by the rate of diffusive transport through the slag or through the newly formed solid phase. Significantly, penetration and thus, corrosion are determined by local slag composition and also by refractory composition and structure. These are also usable handles. The melt properties, particularly viscosity, are an important factor in the corrosion process. Many attack prevention strategies rely on controlling this property by temperature and/or composition manipulations.

Testing to quantify and possibly elucidate the nature of corrosion is an important component of metal–crucible interaction management. In these tests, the extent to which real process conditions are reproduced determines the reliability of the results, which together with thermodynamic calculations can serve to describe the overall process more accurately. An important aspect of testing is the post mortem examination of the failed metal–crucible (and also a successfully operating) system and parts of it. More often, this information complements other tests or provides new directions for evaluation. X-ray, microscopy and metallographic examinations, as well as chemical analysis, are all brought in to complete the picture. In the matter of metal–crucible interactions, focusing on the basics of interaction and using the knowledge thus gained has been a highly effective route for managing the interactions to save the equipment and the melt and to run the process in a cost- and resource-effective manner.

3 Crucible Materials

3.1 INTRODUCTION

Broadly, a crucible can be considered as the material surface at work in contact with the melt and related fluids during material processing. This emphasizes the importance of the popular concept of 'working lining' of process vessels. When a crucible is referred to here, it means a vessel holding the melt and also the lining of a furnace system that is in contact with the metal melt or slag. The work here is containing the melt, resisting and surviving its attack, and withstanding the concurrent thermal and mechanical stresses the process entails; and, doing all this without causing any degradation of the melt. The crucible also needs to maintain its integrity and functionality for durations that are useful in practice. In the field of common metals, both ferrous and common non-ferrous, the crucibles are usually refractory ceramic materials. In the case of less common metals, the crucibles are more specialized and are often made of refractory ceramic compounds, refractory metals and also some iron and nickel alloys.

Crucibles can be made of metals and alloys, oxide minerals, pure oxide compositions, or compounds such as non-oxide ceramics, like metal borides, carbides, nitrides, silicides, phosphides, sulphides and also fluorides. Carbon is an exceedingly useful crucible material, mostly as graphite and also as vitreous carbon. Many of the properties of these materials are available not only when they are used in monolithic form but also when they are used as linings or coatings on other materials. An impressively versatile range of applications is fulfilled by the class known as water-cooled crucibles. Arc and electron beam melting furnaces in a variety of configurations are enabled by water-cooled copper crucibles. While these crucibles impose certain limitations on the processing and on the product, their role in materials processing is unique and irreplaceable. An interesting and highly effective form of water-cooled 'crucible' in induction heating systems is exemplified by skull melting or cold crucible induction melting. This system is made possible by exploitation of the surface tension properties of the melt. The same property is used in another technique that reaches a higher level of sophistication–float zone melting, in both the induction-heated and electron beam-heated modes. Interaction with the crucible is entirely avoided, at least during melting, in the process known as levitation melting. The charge is heated and melted while suspended out of contact with any material container. The amount of material handled by this technique, as of now, is very small. In this chapter, materials that have been used as crucibles are surveyed, focusing on the properties and characteristics that make them relevant for applications and also features that limit their applications.

3.2 CERAMICS

The use of the term *ceramic* implies any of the numerous hard, brittle, heat- and corrosion-resistant non-metallic materials. As a class, ceramics are the most widely

DOI: 10.1201/9780429345562-3

used crucible materials and have been around for millennia. Early civilizations found that certain clay minerals became plastic when sprinkled with water and could be moulded into shapes. They could be hardened by drying in the sun and hardened further in a wood or charcoal fire. Many of the raw materials used by the ancient civilizations are still in use today and are generally known as 'traditional ceramics' (Richerson and Lee 2018).

It is only relatively recently, i.e. during the past few centuries, that naturally occurring minerals have been refined and improved and also new compositions synthesized to achieve enhanced and new properties in ceramics. These improved materials are often called 'modern ceramics'. They are of controlled composition and structure and are often engineered to the needs of applications that have demands far beyond the capabilities of the traditional ceramics. Modern ceramics include plain oxides (such as Al_2O_3, ZrO_2, ThO_2, BeO, MgO and Y_2O_3), oxidic compounds ($MgAl_2O_4$, $BaZrO_3$, etc.), borides, carbides and nitrides (such as TiB_2, TiC, SiC, B_4C, Si_3N_4 and AlN) and many more.

3.3   REFRACTORIES

In day-to-day usage, ceramic structural materials that retain their integrity at temperatures above about 1000°C and under the 'harsh' mechanical and chemical conditions that often accompany such temperatures are known as refractories. A refractory endures not only elevated temperatures but also a certain amount of mechanical load concurrently (Phelps and Wachtman 2011).

The term *high temperatures* is somewhat unspecific but usually means about 1000°C and above, or temperatures at which, because of loss of strength, oxidation or melting, neither the traditional ceramics nor metals can be used. The greatest use of refractories is in the steel industry for construction of linings of equipment such as blast furnaces, converters, transportation ladles and casting equipment. Similar uses abound in non-ferrous metallurgical furnaces and process vessels.

The 'refractoriness' of a ceramicis, in one way or other, rated by the temperature up to which it can resist deformation under load and/or creep. A traditional method is to measure its pyrometric cone equivalent (PCE), and refractory ceramics are formally defined as non-metallic materials having a minimum PCE of 1500°C (Fruehan 1998).

In the beginning, metal melting was the purpose that often started and generally sustained the development of refractories. The refractories were mostly quartz-containing clays. Apart from standing up to modest temperatures, they did not possess many of the properties that modern -day refractory products have by default. Refractory technology was born at the beginning of the 19th century, when the addition of pre-fired clay or grog material was found to result in bricks with better dimensional integrity. The intrinsic properties of a refractory can be significantly enhanced for a given application by suitable techniques for its fabrication into a component.

Refractory materials may be broadly grouped into siliceous, aluminous, aluminosilicates, magnesite, dolomite or chrome ore and inevitably, the miscellaneous category of special refractories. Refractories are used in two basic forms: (i) pre-shaped objects (bricks and shapes) and (ii) unformed compositions or monolithics.

Specialties or monolithics are refractories that are applied to form a mono-lithic, integral structure and cure to a stable dimensional form through hydraulic or chemical setting post application. They are a combination of refractory oxides that include Al_2O_3, SiO_2, Fe_2O_3 and CaO, formulated for specific requirements and capabilities for service at temperatures as high as 1800°C. They are available as castable, mouldable and ramming mixes. The mortars used to install bricks and shapes also belong to this group, and their compositions are chosen to match the brick they bond together. The objective is to achieve a lining that is as close as possible to a uniform refractory structure. Castables, or refractory concretes, are dry, granular refractory mixes suitable for mixing on site with water prior to application. They are used to form complete furnace linings and other non-stan-dard shapes (Fruehan 1998). In hydraulic-bonded castables, the use of MgO is not particularly widespread because of (i) excessive expansion due to hydration when used with cement-containing hydraulic bond, (ii) rapid setting, (iii) high thermal expansion of MgO at steel-making temperature and (iv) low strength at high tem-perature. The use of micro-silica as binder prevents hydration and thus, may avoid the associated problems.

As regards pre-shaped objects, hand moulding, once widely used, is now used only for special shapes, and power pressing is common. Hot pressing and isostatic pressing, both cold and hot, are used for difficult or special refractories. Slip casting is also used for special refractory shapes. Post shaping, the refractories are usually fired in kilns, but not the chemically bonded types. These are pressed and installed without firing. The steel-clad refractories are encased in a metal sheath at the time of pressing. During the heating after installation, iron in the sheath oxidizes and interacts with the refractory and forms a bond between the individual bricks. The common practice with clay refractories is to incorporate calcined clay or grog to increase the density and to reduce the shrinkage on firing.

A high melting point, far above the intended process temperature, is a normal requirement for a crucible refractory. Then, good high-temperature strength, low coefficient of thermal expansion and good thermal shock resistance are all proper-ties that are valuable in refractories. A low propensity to chemical reaction is very important. This property is considered not just in the context of interaction with the melt but also as regards convenience in storage and handling. For example, the tendency of the magnesium oxide or calcia to hydrate is an issue to be circum-vented. The possibility and also rate of corrosion of refractories by molten slags and metallic melts is, of course, a key consideration. When the crucible refrac-tory is porous, corrosion-dissolution often follows penetration into pores and in all similar pathways including interconnected networks of voids. Both physical penetration and chemical invasion are favoured by good melt/slag–solid wetting and by low melt/slag viscosity. Corrosion may start with the dissolution of either the matrix and/or the aggregates. The solid phase may just dissolve into the liquid or may undergo a chemical reaction and cease to exist. Besides, process gas such as CO, SO_2, Cl_2, CH_4, H_2O, and volatile oxides and salts of metals may permeate and take part in the reactions.

Generally, refractories have the ability, within certain limits, to withstand altera-tion by penetration, contamination and/or reaction in service and still function as reliable engineering structures. Many factors control this interaction. Some of these, like porosity and wettability, are manageable to a certain extent. Reference to the thermodynamic data and phase diagrams gives indications as to which combinations of slag and refractory will react. Usually, tests are carried out to validate the indica-tion. Some of the popular tests were mentioned in Chapter 2.

3.4   SILICA-ALUMINA GROUP

Silica (SiO_2) is a major ingredient in many refractories (Fruehan 1998). The raw materials used in the manufacture of conventional silica refractories contain 99% or more SiO_2 and low levels of impurities, particularly alumina and alkalis. Natural silica is the mineral quartz, which is the primary constituent of sand, sandstone and quartzite. Depending on the thermal history, silica refractories may contain propor-tions of quartz, cristobalite and tridymite, which are the different crystalline forms of silica (Figure 3.1).

Fused silica is produced by the actual fusion of very high-grade silica sands in electric arc, electrical resistance or other furnaces. Crystalline raw material is

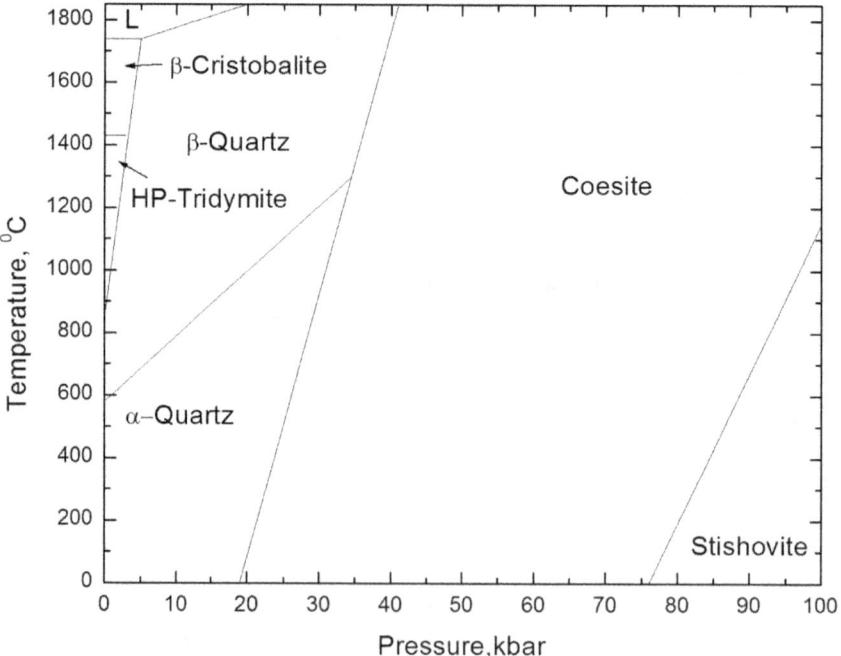

FIGURE 3.1   Phase diagram of silica.

converted into an amorphous glass known as fused silica. The key characteristics of fused silica are very low thermal expansion, low thermal conductivity, high purity and excellent resistance to thermal shock (Fruehan 1998).

The raw materials for silica brick lack a natural bond. Lime is the universally used bonding additive. With additions of CaO to SiO_2, the melting temperature remains unchanged up to about 28% CaO. Silica brick has good tolerance for FeO. But, a small amount (~1%) of Al_2O_3 can ruin this. Silica refractories account for about one-sixth of total refractory production. They are made by firing crushed and ground quartzite (ganister) with about 2% added lime to assist in bonding (Routschka and Granitzki 2005). The outstanding characteristic of silica is its load-bearing ability at high temperatures. It was therefore used in a very wide sprung-arch roof over an open hearth.

3.4.1 FIRECLAYS

In nature, fireclay is found as a normal clay, whiter and lighter in colour but with higher alumina content than normal mud. Fireclays contain the aluminosilicate mineral kaolinite. Fireclay refractories are made from clays containing kaolinite $(Al_2[Si_2O_5][OH]_4)$ with 15–45% alumina and 50–80% silica. Alkalis (~3.5%) and iron oxides (<3%) are the usual impurities. The impurities largely responsible for lowering the refractoriness of fireclay refractories are CaO, MgO, FeO, Na_2O and K_2O. The temperature at which the fireclay is usable (max~1600°C) depends upon the alumina-to-silica ratio and also the impurity content. Fireclays are classified as low-duty, medium-duty, high-duty and super-duty to indicate the intended temperature of use. Firebricks (fireclay bricks) have a relatively low thermal expansion coefficient and thus, moderate thermal shock resistance. They are fairly inert in acidic environments and become reactive in basic environments. A deleterious fluxing agent for fireclay refractories is MnO, a relation similar to that of FeO with the Al_2O_3–SiO_2 system.

3.4.2 ALUMINOSILICATES

Fireclay is the raw material from which the bulk (about 70%) of aluminosilicate refractories are manufactured. High-alumina refractories are made from alumina-rich bauxitic kaolins or bauxite clay combinations. These materials are mined, blended and fired in rotary kilns to yield calcined aluminosilicates from about 50% to over 70% alumina content (Fruehan 1998).

The refractoriness of alumina–silica refractories will be seriously affected by very small (<1%) amounts of Na_2O or K_2O (Fruehan1998). Figure 3.2 shows the Al_2O_3–SiO_2 phase diagram. Impurities generally present in most commercial refractories in this system can have a pronounced effect on the usable service temperatures vis-à-vis those indicated in the phase diagram. In the FeO–Al_2O_3–SiO_2 ternary, the liquid can form even below 1095°C. Iron oxide-bearing liquids are characteristically very fluid and have high degradation potential.

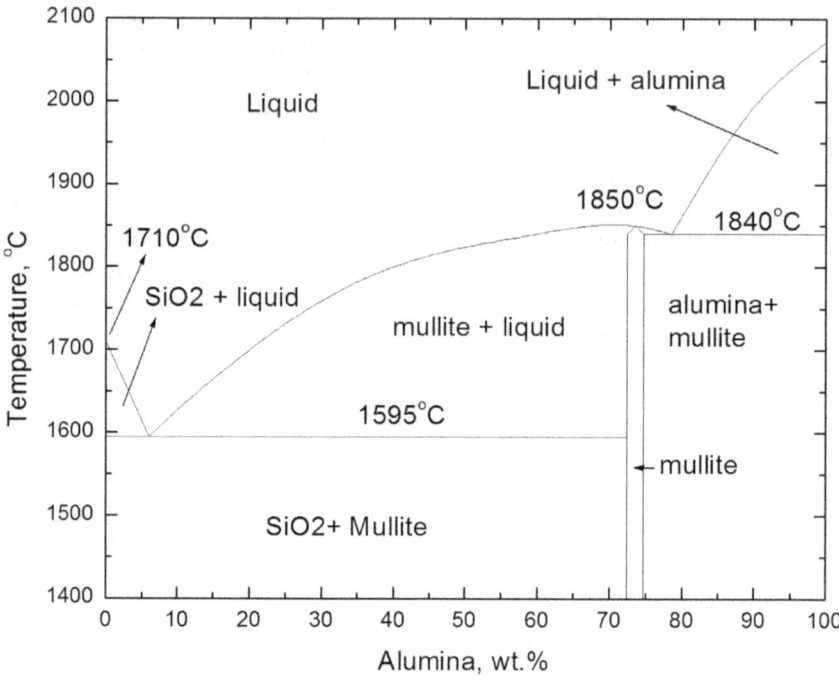

FIGURE 3.2 Alumina–silica phase diagram.

3.4.3 MULLITE

Mullite rarely occurs in nature. It is produced commercially by firing various alumi-nosilicates. Andalusite is considered the best precursor. The general composition is $3Al_2O_3.2SiO_2$. It has a good combination of properties for a refractory: high thermal stability, low thermal expansion, high bending strength and creep resistance, and excellent resistance to thermal shock. The properties of aluminosilicate refractories are largely determined by the presence of two crystalline phases, mullite and corun-dum, and the siliceous glassy phase. In the absence of glassy phase formation in the grain boundary, mullite can retain >90% of its room-temperature strength at more than 1600°C.

High-alumina bricks are produced using natural raw materials, high-grade (>45% Al_2O_3) clay, bauxite and the sillimanite group of minerals, as well as synthetic mate-rials (fused corundum, mullite, calcined alumina and sintered corundum). Bauxite is primarily used for the manufacture of 76–85 wt% alumina-containing high-alumina refractories. To develop the mullite matrix (to improve thermal shock resistance and other high-temperature properties), materials from the sillimanite group with cal-cined alumina are added. In aluminium melting furnaces, bauxite bricks are used in the chemically bonded form.

3.4.4 ALUMINA

Aluminium oxide (Al_2O_3) occurs naturally as the mineral corundum. Alumina powder, used to manufacture polycrystalline Al_2O_3 ceramics, is produced from the mineral bauxite by the Bayer process. Bauxite is primarily colloidal aluminium hydroxide intimately mixed with iron hydroxide and other impurities.

Many types of chemically and thermally processed alumina are used in refractories. These are calcined, tabular and fused alumina, all obtained from the alumina made by the Bayer process. Calcined alumina is used to promote refractory binding during manufacture or use, whereas tabular or fused products form very stable aggregates. Tabular alumina is formed by calcination at 1925°C, whereas the fused alumina, which is denser, is made by total melting and rapid solidification (Fruehan1998).

Tabular alumina has excellent high-temperature properties and thermal shock resistance. Its name is derived from the large (typically 100 to >400 μm), flat, table-like α-alumina crystals in the microstructure. Tabular alumina is different from white fused alumina. The presence of large numbers of closed pores and controlled pore size makes tabular alumina superior in thermal shock resistance to white fused alumina and other aluminosilicates.

3.5   MAGNESIA–LIME GROUP

The magnesia or magnesia–lime group encompasses all refractories made from dolomite and synthetic magnesites. These are the principal refractories used for basic steelmaking. Magnesia is noted for its compatibility with iron oxides and therefore forms the base for all types of basic refractories, including those made from magnesite, olivine, dead-burned dolomite, and magnesite and chrome ore. Calcia is more reactive with iron oxide, forming low-melting calcium ferrites such as dicalcium ferrite ($2CaO \cdot Fe_2O_3$), which melts incongruently at~1440°C. Calcia is more prone to hydration and disintegration.

3.5.1 MAGNESIA

Periclase (MgO) is an extremely rare mineral and does not occur in commercial quantities. About three -fourths of the world's magnesia production comes from magnesium-rich minerals. Magnesite ($MgCO_3$) is the most important, followed by dolomite (CaMg ($CO_3)_2$). Deep well brines and sea water have also been important sources of magnesia (Canterford 1985). Magnesium hydroxide, $Mg(OH)_2$, is precipitated from brines (deep well type) and seawater by reaction with calcined dolomite or limestone and treated in a rotary kiln to produce refractory-grade magnesia.

Commercial magnesia is made in three basic forms: caustic-calcined, dead-burned and fused. Caustic-calcined magnesia has a high surface with a small (1–20 μm) grain size and is reactive. Dead-burned magnesia (grain size 30–120 μm) is formed at temperatures above 1200°C. The calcined magnesia is briquetted and fired in shaft kilns at temperatures up to 2000°C. The product is sintered, dense refractory-grade magnesia. By melting a refractory-grade magnesia or other magnesia precursor in an electric arc furnace, fused magnesia (grain size 800 to 1000 μm) is made. The molten mass from the furnace is cooled and broken up for use in refractories (Fruehan1998). For magnesia (MgO), the maximum operating temperature is 1800°C depending on grade. Magnesia ceramics usually contain around 90% MgO together with Al_2O_3–SiO_2 and CaO. The magnesia spinel variety contains around 90% MgO together with Al_2O_3.

3.5.1.1 Magnesite Bricks

Commercial-grade MgO brick contains atleast 80 wt% of MgO (Biswas and Sarkar 2020). Magnesite has poor sinterability, which is improved by adding bauxite. Magnesia–alumina spinel ($MgAl_2O_4$) and forsterite (Mg_2SiO_4) form during firing. In unfired bricks and bricks clad in steel casing, in situ sintering occurs during heat up, and properties similar to those of fired magnesite bricks are developed.

In magnesite raw material, the impurities present include SiO_2, Al_2O_3, Fe_2O_3 and B_2O_4. Their level must be kept low to preserve the refractory properties. The properties of the magnesite bricks strongly depend on the CaO/SiO_2 ratio. The coefficient of thermal expansion of pure MgO refractory is very high. Consequently, the thermal shock resistance is poor, but it can be improved using additives.

3.5.2 DOLOMITE

The double carbonate dolomite ($CaCO_3 \cdot MgCO_3$) occurs in nature. It can be converted to refractory dolomite (CaO·MgO) by firing. Silica, iron oxide and alumina are the persistent impurities. Manganese, sulphur and phosphorus are also present (Fruehan 1998). Refractory-grade dolomites are usually enriched in magnesia, containing a minimum of 20% MgO and a maximum of 2.5% impurities.

Dolomite bricks are made of sintered dolomite and fused dolomite. They consist of around 60% CaO and 40% MgO. In dolomite, CaO and MgO remain as a eutectic mixture of solid solutions, as shown in the phase diagram (Figure 3.3). MgO grains are embedded in a fine matrix of CaO, and the matrix has a broad composition range. Hence, oxides such as MgO, TiO_2 and ZrO_2can be readily incorporated in the matrix. Fired direct-bonded dolomite bricks are made by mixing graded dolomite with organic binders, compacting and firing at high temperature. To improve the corrosion and hydration resistance, the fired bricks are impregnated with tar or pitch.

Doloma brick shows a high degree of direct bonding. Direct-bonded doloma bricks are susceptible to thermal shock, which is significantly improved through the addition of small amounts of zirconia. Improvements to the slag resistance of both direct-bonded and carbon-bonded doloma bricks have been made through the addition of high-purity magnesia. Tar-bonded dolomite has been in use for some time. Thermosetting phenol–formaldehyde resin is presently used instead of tar/pitch. A chemically bonded dolomite brick with 3–4% residual carbon is less wetted by slag and has better thermal shock resistance.

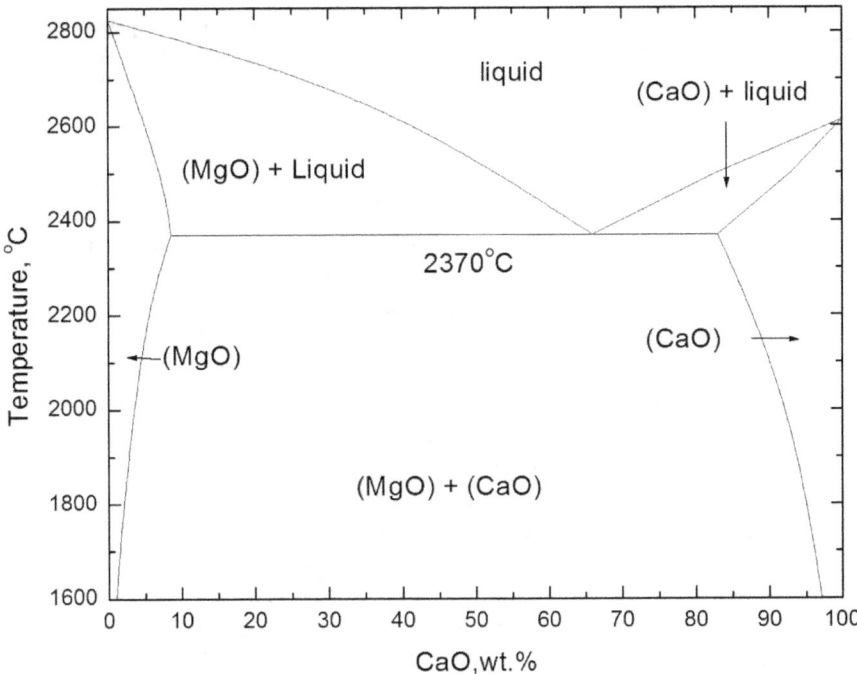

FIGURE 3.3   Calcia–magnesia phase diagram.

3.5.3   MAGNESIA–CHROME

Mag–chrome or $MgO–Cr_2O_3$ refractories find predominant use in copper smelting, converting and refining furnaces (Schlesinger 1996). Widespread use of mag–chrome began in the 1950s and 1960s and continues to date. Chromite ore contains CrO and MgO and also Al_2O_3, Fe_2O_3 and SiO_2 in significant amounts. The first mag–chrome refractories produced, and which still remain in use, are the silicate-based ("burned") materials. They are obtained when the chromite ore is fired at temperatures <1500°C with or without added magnesia. Their microstructure shows grains of a spinel-based structure, $(Mg,Fe)(Al,Fe,Cr)_2O_4$, surrounded by a rim of fosterite Mg_2SiO_4 (m.p. ~1900°C) and silicates. The silicate phases are low melting (~1200°C for fayalite), and the usefulness of such refractories is thus limited.

Direct-bonded refractories were first produced in the 1950s and possess superior high-temperature properties. Firing mixtures of chrome ore and magnesite at 1600–1700°C results in the formation of a liquid silicate phase. On cooling, precipitation of secondary spinel from the silicate phase facilitates bonding by interdiffusion between the spinel grains themselves. To further enhance the properties of direct bonded refractories beneficiated of chrome ores, that have lower silica content are used as start materials

Fused cast mag–chrome refractories are produced by melting the entire mixture in an electric furnace at about 2450°C and casting the melt. The fused grain materials still contain some silicate phase as minor discrete particles in contrast to the continuous phase in the burned and direct-bonded bricks. Mag–chrome refractories produced from purer materials perform better. A higher calcia/silica ratio in the mag–chrome refractory promotes the presence of more refractory silicate compounds, and the refractory has higher temperature capability. A higher Cr_2O_3 content relative to the combined amount of Al_2O_3 and Fe_2O_3 has been associated with greater resistance to silicate slag attack.

Magnesia and chrome ore synergism results in excellent properties in the refractory. One constituent tends to minimize the major weaknesses of the other constituent. The refractory is called chrome–magnesite or magnesite–chrome, denoting that the first named is the dominant constituent. The properties of the bricks are a function of the magnesite-to-chrome ratio, among other things.

Rebonded magnesia–chrome brick is also made with synthetic fused magnesia, shaped under high pressure and fired at very high temperature. The brick structure then consists of large, fused aggregates embedded in a matrix of finer fused grains. Another method was first to electrically fuse the chrome ore and the magnesia grains at about 2450°C, pour the melt into moulds, cool, crush, add binder, press and fire at 1550°C to manufacture the bricks. Electrofused magnesio-chromite is a mixture of chromite crystals inside a slag glass mainly composed of MgO (Bravo et al. 2017). Chrome–alumina is another refractory, mainly used in induction crucibles. The typical composition of the ceramic is Al_2O_3 92.8%–SiO_2 2.6%–Cr 4.0%+trace oxides.

3.5.3.1 $MgO–Al_2O_3$

The $MgO–Al_2O_3$ phase diagram is similar to that of $MgO–Cr_2O_3$ (Schlesinger 1996). Refractories based on mixtures of MgO and Al_2O_3 have been considered for replacement of mag–chrome. Periclase and alumina rarely occur combined in nature, and $MgO–Al_2O_3$ refractories are made from purified raw materials. Castable and monolithic magnesia–alumina materials are produced using a binder (calcium aluminate cement) or by direct bonding. Fused cast materials are also made.

3.5.4 MAGNESIA–CARBON

The incorporation of carbon in magnesia refractories results in much improved performance in terms of higher thermal conductivity, good thermal shock resistance and resistance to slag penetration, and better high-temperature strength and refractoriness. Although the magnesia–carbon refractories are chemically quite simple, their properties may be extensively changed by varying the carbon level (or C/MgO ratio), type of carbon, type of magnesia and type of bond (Mikami and Martinet 1979; Canterford 1985). Carbon is normally introduced by tar-bonding, pitch impregnation and graphite additions.

The magnesia–carbon bricks are further grouped as tar-impregnated fired magnesia bricks (2–4% C), carbon-containing impregnated magnesia bricks (7% C), and resin-bonded MgO–C bricks (20–25%). Sintered and fused magnesia with larger crystal size is used as the source of MgO. Graphite is normally the source of carbon; other sources of carbon are carbon black, hard pitch and powder resins. Other things being equal, purer MgO and graphite have improved physico-chemical properties and corrosion resistance at high operating temperature. Antioxidants, such as Si, Al, metal powders and B_4C, are also added to the mixture to improve oxidation resistance and high-temperature strength. The functioning of MgO–C refractories is determined by the direct oxidation of carbon (and graphite) by air; the reduction of magnesia (and other oxides) by carbon (graphite); and the role of antioxidants. These are covered later in this chapter in the section on corrosion.

Magnesia–alumina–graphite refractories are Al_2O_3– bricks that have MgO as additive (Biswas and Sarkar 2020). When used in ladle operating conditions, the spinel $MgAl_2O_4$ forms with an volume expansion of (thermal). This expansion closes the joints of the bricks and ends up reducing metal penetration. With higher magnesia content, the corrosion resistance is improved due to the formation of spinel, but the larger thermal expansion here may end up cracking the bricks. Al_2O_3–MgO–C (AMC) emerged as an alternative working lining refractory to MgO–C in the 1980s. The performance of AMC bricks depends on the alumina aggregates used. In the ladle bottom, high-quality AMC bricks made of tabular alumina are used.

3.6   CARBON

Carbon is the only elemental refractory that is also a non-metal. It is unmeltable and is not wetted by metal melts, slags or salts under oxygen-free conditions. It dissolves in pig iron melt and to a lesser but significant extent in certain reactive and refractory metals. It has high thermal conductivity and low thermal expansion coefficient, and this combination imparts to it high thermal shock resistance. The major limitation of graphite is its susceptibility to attack by oxygen, steam and CO_2 above 400°C.

Much of the graphite used in industry is synthesized. The raw materials (low ash cokes, anthracite, and natural and artificial graphite) are blended together with tar or pitch in heated mixers, and the hot mix is compacted by ramming or vibration. The cold, hardened shapes are first fired at temperatures >400°C in the absence of air and are eventually converted to graphite by self resistance heating at ca. 3000°C. Depending on the quality and properties of the various raw materials used, graphite with widely varying properties (thermal conductivity and oxidation, erosion and alkali resistance) may be obtained.

In metallurgical processing, the principal use of graphite is as large carbon blocks in the bottom and bosh of a blast furnace (iron), or as electrode cum lining for aluminium electrolysis cells (Phelps and Wachtman 2011).

Generally, graphite is used in refractories to reduce the wetting of the refractory by the melt and/or slag and to increase the thermal conductivity, which will result in better thermal shock resistance of the refractory. In oxide–carbon refractories, the range of carbon content used is from 4% to 35%. With the increase in graphitic content, the thermal conductivity of the refractory increases, but there is a decrease in the density of the refractory, not only because graphite is much lighter but also because the graphite flakes are not very conducive to close packing, unlike the granules (Fruehan 1998).

Refractory bricks made of carbon can be classified into three groups: amorphous, semi-graphite and graphite blocks. In recent years, two other superior-quality carbon blocks have become available: micropore and super micropore. The average pore diameter in these varieties is below 1 μm, and the thermal conductivity is also high.

3.7 BERYLLIA

Chemically, BeO is one of the most stable materials. In many of its properties – thermal expansion, and electrical and mechanical properties – BeO is similar to alumina. Thermal conductivity is relatively high, and for a material of good strength, this gives better thermal shock resistance than Al_2O_3. Beryllia crucibles can be excellent containers for liquid metals. One limitation is that it reacts with water vapour at high temperatures to form a stable hydroxide species, probably gaseous $Be(OH)_2$, which causes it to evaporate significantly in a moist atmosphere (Brook et al. 1991).

Beryllia powder is commercially available with purities over 99% BeO. It can be readily fabricated to high-density (<5% porosity) shapes by normal ceramics fabrication methods such as pressing, extrusion, injection moulding or slip casting. Sintering temperatures are in the range 1600–1800°C. Hot pressing in graphite dies and hot isostatic pressing may also be used. Hot pressing, as well as slip casting and sintering, has been used to achieve near theoretical density (<1% porosity) with the purest powder.

Beryllia is toxic. This has led to increased costs of production and inhibited use. Inhalation of fine beryllia dust results in respiratory disease in an acute form from short, concentrated exposures or in a chronic form from long, continued exposure. Special precautions are required when handling fine powder.

3.8 ZIRCON AND ZIRCONIA

Zircon, or zirconium silicate ($ZrO_2.SiO_2$), is a naturally occurring mineral having excellent refractoriness (Fruehan1998). The main raw material for zircon brick is zircon sand of 100–300 μm particle size. Zircon bricks have good wear resistance, less wettability by molten metal and slag, and hence, good corrosion resistance. Alumina is added to zircon to form zircon–mullite compositions, which are more resistant to thermal shock and corrosion than plain zircon bricks. In steel making, zircon refractory is used in the tundish nozzle due to its non-wettability by liquid steel and good resistance to erosion and spalling.

Zirconia has a combination of high melting point (2715°C) and potential for high strength with a maximum operating temperature of 2400°C depending on grade (Fruehan 1998). Zirconium oxide (ZrO_2) occurs naturally as the mineral baddeleyite, commercially used zirconia is generally obtained by processing zircon. Combined with high melting point, the superior resistance to corrosion and erosion makes zirconia an attractive refractory (Bullock et al. 1989). Zirconia is polymorphic. The stable room-temperature phase is monoclinic. Around 1170°C, it transforms to the tetragonal, which is stable up to 2370°C, when it becomes cubic. The tetragonal/monoclinic transformation is associated with a large volume change (~4%). This is

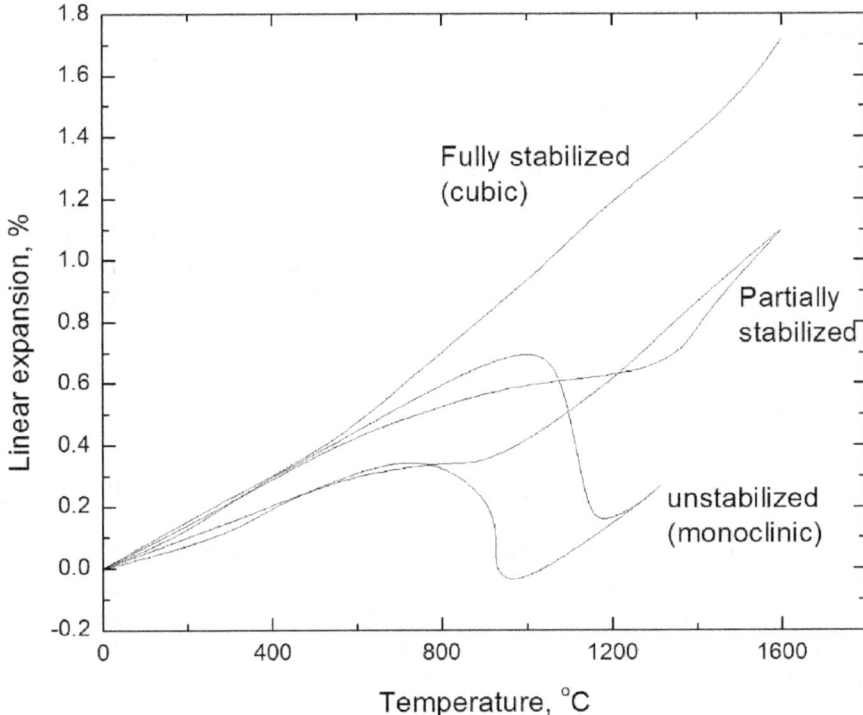

FIGURE 3.4 Thermal expansion behaviour of zirconia (From Routschka, G. (Ed.). 2001).

the reason why pure zirconia wares break as their temperature changes through this value. To counteract the deleterious effect of phase changes, zirconia may be converted fully to the cubic phase by incorporating small amounts of calcia, magnesia or yttria. The cubic phase then remains stable at all temperatures. Stability, thermal shock resistance and hot load properties are all enhanced in the stabilized zirconia (Fruehan 1998).

Fully stabilized zirconia refractories have very low thermal shock resistance because of their high thermal expansion coefficient (Figure 3.4) and low thermal conductivity. In many applications, however, good or even excellent thermal shock resistance is required. With zirconia, this is achieved by optimizing the level of stabilization (Johns and King 1970). Crucibles for induction melting of refractory metals are made with zirconia compositions having a cubic zirconia content of about 40%. In tundish nozzles in the casting of molten steel, a cubic zirconia content of 65% is considered suitable. In the arrangement for partial stabilization, cubic materials contain fine dispersions of tetragonal phase. Crack propagation is hindered when the metastable precipitate transforms, resulting in significant improvements in toughness and strength. Some toughening in zirconias can also result from the presence of microcracks. This flexibility, combined with the capability of microstructural control through heat treatment, is the reason for the popularity of zirconia, as it can be engineered to fit a wide variety of special refractory applications.

Zirconia is used in the natural, stabilized or fused state, the last often as a mixture with alumina, silica or other compatible oxides (Fruehan 1998). The manufacture of zirconia products starts with stabilized or partially stabilized zirconia raw materials, which are ground and mixed, and the graded material is mixed with organic binder, shaped in a press, dried and fired at >1700°C. Zircon with alumina can be processed to form zirconia–mullite grains. This material has high thermal shock resistance as well as good corrosion resistance. The well-known use of this material, with the addition of carbon, is in manufacturing alumina–zirconia–carbon slide gate plate refractory in steel ladles.

3.9 YTTRIA

The rare earth oxide yttria is one of the most stable refractory ceramic oxides. Its availability, in high purity, improved with the expansion of the rare earths industry, particularly after the late 1970s.

Yttria crucibles can be made by dry pressing followed by sintering. A crucible with a somewhat porous structure is considered conducive to good thermal shock resistance. Such porous crucibles are usually fabricated from a mixture of ceramic raw material powders with varied diameters ranging from coarse to fine (Tetsui et al. 2012a). Other normal ceramic fabrication methods have also been used. The main limitations of yttria are its cost and modest thermal shock resistance.

3.10 SPINELS

Spinel refers to a group of minerals of the common structure AB_2O_4, where A represents a divalent cation and B a trivalent cation. The common trivalent cations are aluminium (Al^{+3}), ferric iron (Fe^{+3}) and chromium (Cr^{+3}), and the divalent cations are magnesium (Mg^{+2}) and ferrous iron (Fe^{+2}).

Magnesia–alumina spinel ($MgO \cdot Al_2O_3$) and magnesia–chrome spinel are useful refractories. The only compound in the binary system $MgO–Al_2O_3$ is magnesium–aluminate spinel, with a formal stoichiometric ratio of 71.8% Al_2O_3–28.2% MgO. $MgAl_2O_4$ (MA) spinel has a range of stoichiometry. It has a cubic crystal structure. Magnesium–aluminate spinel is a highly refractory material, with a melting point of 2135°C, low thermal expansion, good thermal conductivity and high thermal shock resistance. All types of spinel tend to form substitutional solid solution swith each other and with other oxides. In $MgO \cdot Al_2O_3$ spinel, both magnesium and aluminium cations can be replaced by other cations of similar size. This results in the formation of alumina-enriched spinel as well as magnesia-enriched spinel. Alumina-enriched spinel can absorb FeO or MnO by substitution of MgO and consequently retard the infiltration of slag into spinel-containing refractory. The picking up of Fe^{2+} and Mn^{2+} from the slag into the spinel leads to an expansion of the lattice and effectively prevents further penetration of ladle slags. When the slag penetration of alumina, spinel and aluminaspinel castables was compared, castables with 20% spinel displayed the lowest penetration and corrosion, the reason being that alumina and spinel in the castable react with CaO, FeO and MnO in the slag, locally increasing its melting point and viscosity. This type of complex interaction, involving the pick-up of rapidly diffusing cations from the slag into solid phases (such as Fe and Mn into spinel) and dissolution of solid (initially fine, reactive matrix phases) into the slag, altering its local composition, has also been highlighted by Goto et al. (1997) and Korgul et al. (1997). The liquid product of this interaction is the phase in contact with the solid phases of the refractory and determines any penetration and corrosion.

$MgAl_2O_3$ spinel can be introduced into a refractory composition by forming it in situ or by the addition of pre-formed spinel. Spinel is formed in situ by adding magnesia to high-alumina refractory or adding alumina to magnesia-enriched refractory. During the firing of bricks or during preheating, MgO reacts with Al_2O_3 to form the spinel. Spinel formation is accompanied by volume expansion. Pre-formed and in situ spinel-containing refractory are very similar with respect to chemical composition and bulk density of the product, but they do show distinct differences in other physical properties, such as thermal expansion, hot strength and slag corrosion resistance.

Pre-reacted spinel does not present an expansion problem. This can be added in larger quantity to high-alumina refractory. Expansion during in situ spinel formation can result in cracking of the refractory. However, in situ spinel formation in refractory bricks and castables is advantageous for high-temperature strength and thermal shock resistance (Biswas and Sarkar 2020). Hence, addition of MgO to high-alumina refractory to form in situ spinel is restricted to about 3–5% MgO to form 12 to 18% spinel, and pre-formed spinel is added as necessary to reach the desired spinel content for high corrosion resistance. Refractories containing MA spinel are potential replacements for Cr-containing refractory linings in steelmaking (Lee and Zhang 2004) on account of their excellent thermal shock resistance and refractoriness

As an alternative to other carbon-free refractories, alumina spinel fired bricks show superior performance in the highly corrosive environment in secondary steel making (Biswas and Sarkar 2020). The alumina spinel bricks are made using high-purity synthetic alumina aggregates, white tabular alumina and white fused alumina and $MgO–Al_2O_3$ spinel, MgO and calcined alumina in finer fractions. High-alumina bricks with spinel addition also have the advantages of lower thermal conductivity and low thermal expansion coefficient compared with MgO–C or Al_2O_3–MgO–C (AMC). The spinel-containing alumina being carbon free, dissolution of carbon in ultra-low-carbon steel is avoided. This type of brick is widely used in producing clean steel. Pre-reacted spinels have good thermal shock behaviour. This is mainly due to differences in thermal expansion between alumina and spinel. The differential expansion leads to microcracks in the matrix that act as crack arresters.

3.11 CARBON-CONTAINING REFRACTORIES

Carbon-containing refractories are widely used in the steel industry. Carbon is used in two forms: pyrolytic carbon derived from pitch or resin binders, which ends up in the matrix in fine and reactive form, or graphite flakes, which may be considered as part of the aggregates. Carbon in both forms is used with many refractory oxides such as Al_2O_3, SiO_2, ZrO_2 and MgO. A typical unfired microstructure contains large sintered and fused MgO grains bonded to graphite flakes by a nanoscale carbon bond. MgO–C refractories are used in basic oxygen steel-making (BOS) and electric arc furnace (EAF) steel-making furnace linings.

While carbon confers many of the good properties on oxide–C refractories, it is also prone to oxidation. The weak nature of the C bond between the oxide and graphite lowers the strength of the refractory. To inhibit oxidation and provide stronger (ceramic) matrix bonding at high temperatures, metallic or carbide powder is added. Several mechanisms for the good slag resistance of C-bonded doloma and MgO refractories have been proposed (summarized by Lee and Rainforth 1994). These include carbon's non-wettability, so that liquid penetration is limited; reduction of MgO by C to form magnesium vapour, which transports, re-oxidizes and deposits as a dense layer of secondary MgO at the slag–brick interface, physically preventing penetration; reduction of iron oxides by C to Fe metal, blunting their ability to attack the oxides and form low-melting phases; and formation of CO vapour by oxidation of C and/or Mg vapour (from reduction of MgO), which provides an overpressure resisting slag/metal ingress. Which one of these mechanisms is the most significant remains an unresolved issue (Lee and Zhang 2004).

The slag corrosion of oxide–C refractories follows the sequence: (i) formation of a decarburized layer due to oxidation of graphite/pyrolytic C by oxygen in the furnace atmosphere or FeO in the slag, (ii) infiltration of slag into the decarburized layer and erosion of the oxide grain by slag penetration, aided by softening of any intragranular silicate bond, and (iii) reaction and dissolution of the oxide grain into slag and molten steel at the prevailing high temperatures (Lee and Zhang 1999).

The protective role of a secondary MgO layer formed at the hot face of MgO–C bricks has often been highlighted (Lee and Zhang 2004). In both dolomite and MgO refractories, this arises from reduction of MgO by C to Mg vapour, which is transported to the hot face and re-oxidized to a fine-grained, dense MgO layer on the hot face. This layer prevents slag/metal ingress into the refractory. This dense layer does not always form. Specific conditions both inside and outside the refractory are required to maintain it; a high CaO/SiO_2 ratio in the slag is required, which maintains the delicate balance between dense layer dissolution and formation by limiting the former. To improve the corrosion resistance of the refractory, MgO purity should be higher, grain size larger and residual impurities SiO_2 and B_2O_3 lower. Carbon in MgO–C refractories is present as graphite or pyrolysed carbon from pitch and resin. The advantage of one form over the other is yet to be established.

Two types of additives mainly used in oxide–C refractories are metals and alloys (such as Al, Si, Mg and Al–Mg) and boron-based compounds (such as B, B_4C, CaB_6, ZrB_2, $Mg_3B_2O_6$ and SiB_6). These two types work in different ways. Metal/alloy additives act by inhibiting carbon oxidation and forming high-temperature ceramic bonds, thus improving hot strength. Metal additives facilitate dense layer formation both at the slag–refractory interface and locally inside the brick, leading to early protection in service (Lee and Zhang 2004). Slag composition also has a significant effect on the corrosion resistance of MgO–C refractories. In general, corrosion rate increases with increasing Cr_2O_3, iron oxide and MnO but decreases with increasing MgO or basicity in the slag.

All said, Lee and Zhang (2004) identified two types of attack on refractories by silicate liquids: direct, wherein atoms of the solid dissolve directly into the liquid, and indirect, in which a boundary layer forms between the original solid and liquid. Direct attack is usually undesirable, while the indirect mode may be beneficial if the boundary layer turns protective. As often highlighted, the influence of the local liquid at the point of penetration into the refractory has a critical effect on the attack. Its composition is influenced by the bulk liquid and/or the refractory. Rapidly diffusing species in the bulk liquid, such as Mn and Fe, migrate to the local liquid. The refractory may take up species from the local liquid or release its own, thereby altering the local composition and hence, viscosity. The phases in the refractory may also dissolve in the local liquid, altering its composition compared with bulk liquid. Slag attack can be controlled through the refractory, whose composition is arranged to make the local liquid viscous by uptake or release of suitable species, or by development of a tight texture and/or dense layers at the liquid/refractory interface. Alternatively, the slag composition can be controlled to minimize attack on the refractory by e.g. saturating it in the main refractory phases. In practice, passive, protective coatings can be generated by alternative process techniques such as slag splashing in BOF vessels.

The function of the slag in metal winning operations is to provide a vehicle for impurity removal initially from the ores and later from the as-smelted molten metal. Normally, a slag attacks the refractory. But, the attack is managed in many ways.

A novel new use of slag is as an in situ refractory. This is generally similar to the formation and functioning of a frozen skull in water-cooled crucibles, common in the handling of very reactive metals. In situ refractory is routinely created in the slag splashing operations in BOS vessels. Lee and Moore (1998) outlined four types of in situ refractories: (i) those arising due to reactions solely within the components of the brick or monolith without any external contribution, e.g. formation of a MgO layer initiated by reaction between carbon and magnesia in the refractory; (ii) those in which reactions occur within the refractory but may be assisted by reaction with the (liquid or vapour) furnace contents, e.g. magnetite formation in the upper regions of flash smelting furnaces; (iii) refractories that react with the furnace contents, generating a protective interlayer between the refractory and furnace contents. Often, the interlayer forms as a result of indirect corrosion of the refractory solid by penetrating liquid, e.g. the formation of a corundum layer on aluminosilicate refractories in aluminium melting and holding furnaces; and (iv) those in which the slag from an earlier batch is splashed or otherwise deposited on the interior of the furnace lining. The next batch process fluids would have this slag deposit as the contact zone.

3.12   BARIUM ZIRCONATE

For the use of a ceramic as a crucible material, it is essential that the compound shows no crystallographic phase transitions in the temperature region of interest. This is in fact the case for $BaZrO_3$. Barium zirconate has a cubic perovskite structure and a high density of 6.242 g/cm^3 (Erb et al. 1995). It has a melting point of about 2600°C, a smaller thermal expansion coefficient than ZrO_2 and therefore, better thermal shock resistance.

In a binary ceramic such as barium zirconate, any deviation from the stoichiometric composition will lead to precipitation of the off-stoichiometric surplus at the grain boundaries of the ceramic. This region will be prone to corrosion.

$BaZrO_3$ powder is synthesized by heating $BaCO_3$ and ZrO_2 (purity >99%) in air at 1200°C and fabricated by cold isostatic pressing (CIP). On CIP of the pre-reacted powders (0.5–1 μm) at about 2000 bar, followed by sintering in air at about 1700°C for 48 h, ceramic shapes with up to 98.5% theoretical density (TD) is obtained. A Ca-doped $BaZrO_3$ crucible could be made starting from $BaCO_3$, ZrO_2 and CaO in the required mole ratios. The Ca-doped $BaZrO_3$ refractory consists of $Ba_{1-x}Ca_xZrO_3$ and CaO phases. Similarly, the yttria-containing variety is also obtained. However, doping with Y_2O_3 decreases the sinterability of the $BaZrO_3$ crucible (Chen et al. 2017).

3.13   CALCIUM ZIRCONATE

Calcium zirconate ($CaZrO_3$) is a stoichiometric phase with a perovskite crystal structure in the pseudo binary $CaO-ZrO_2$ system. It has a high melting point of 2368°C and a chemical stability as good as CaO (Schaffoner et al. 2018).

The zirconate, unlike CaO, is inert to hydration. $CaZrO_3$ can be prepared by the solid-state reaction (between 700 and 1300°C) of equimolar $CaO-ZrO_2$ mixtures with subsequent CIP and sintering (Li et al. 2010) at 1750°C for 3 hours. The crucible was used to melt NiTi and Ti6Al4V alloys at 1500 and 1750°C, respectively, without contamination.

Coarse-grain $CaZrO_3$ has also been prepared by slip casting of $CaCO_3$ and ZrO_2, firing at 1500°C for 3 h and crushing. The product had 12.4% open porosity. Schaffoner et al. (2013) used both solid-state synthesis and electric arc melting methods to prepare coarse-grained calcium zirconate. The fused calcium zirconate had considerably lower porosity than solid-state synthesized material, but during electric arc melting, the CaO/ZrO_2 ratio decreased from 1 in the start material to less than 1, resulting in the formation of cubic ZrO_2 as a separate phase. After crushing and milling of fused $CaZrO_3$, the preparation of large crucibles by CIP resulted in crucibles with porosity of 15.9–16.5%.

Schaffoner et al. (2018) also prepared $CaZrO_3$ refractory concrete (castable) with calcium aluminate cement as a binder in the form of medium-sized crucibles. Although the raw materials contained only a small amount of silica in bulk, a large amount was accumulated in the calcium aluminate matrix, suggesting that the matrix acted as a sink for the silicon oxide impurities of all raw materials.

Calcium aluminate binders were found to enable fast processing and sufficient refractoriness for medium-sized melting crucibles. The properties of uniaxially pressed and cast crucibles of $CaZrO_3$ were generally similar. However, the matrix containing calcium aluminates undergoes pronounced sintering shrinkage, resulting in smaller apparent porosity and a lower refractoriness compared with uniaxially pressed refractories.

The calcium aluminate cement consisted mainly of $CaAl_2O_4$, while it also contained minor amounts of $CaAl_4O_7$ (Braulio et al. 2011). As in other components, impurities should be avoided in calcium aluminate, since they impair not only the refractoriness but also the corrosion resistance.

3.14 SILICON CARBIDE

Special refractories include many types of materials, and silicon carbide is one of them. Silicon carbide, SiC, has good thermal and electrical conductivity, good thermal shock resistance and strength at high temperatures. SiC has been found occurring naturally only as small green hexagonal plates in meteoric iron. This same hexagonal polymorph (α-SiC) has been synthesized commercially in large quantities by the Acheson process (Richerson and Lee 2018). This simple process simultaneously produces lower-grade SiC for abrasives and high-grade SiC for electrical applications. The Acheson process consists of passing electric current through a large, elongated mound of SiO_2 sand mixture, thus heating it to about 2200°C. After cooling, the mound is broken up and sorted to get various grades of SiC.

SiC can be prepared from almost any source of silicon and carbon. Silicon carbide bricks find increasing use in the blast furnace lining (in the shaft). They resist corrosion by acid slags, alkali vapour, zinc vapour and oxidizing gases. The first silicon carbide refractories were bonded with clay, and the refractory properties of the bond placed the limit on the material performance. Self-bonded silicon carbide has removed this limitation. The bonding phase can be SiC itself or silicon nitride or sialon. Silicon carbide crucibles are used extensively for gold and silver melting. Crucibles are of typical composition SiC 67.9%–Al_2O_3 23.4%–SiO_2 8.3% with some oxides in trace levels. They have excellent thermal shock resistance and can be used up to 1500°C.

3.15 BORIDES

Diborides, especially TiB_2 and ZrB_2, are chemically stable in the environment of a Hall–Héroult cell for aluminium production. They are not wetted by molten aluminium, and their electrical conductivity is high (Brook et al. 1991). In aluminium electrolysis cells, replacing the conventional graphite cathode with diborides has often been seriously considered.

Diborides can be used as crucibles for molten metals such as aluminium, copper, silver, gold, zinc, cadmium, magnesium, germanium, lead, bismuth and chromium. Many tonnes of TiB_2 powder are consumed for the production of vacuum metallizing boats (TiB_2, AlN, BN composite). They are resistance heated to about 1427°C and used for the continuous evaporation of aluminium to produce thin metal films.

3.16 BORON NITRIDE

Boron nitride in its hexagonal, soft modification finds use up to very high (~2000°C) temperatures. It is an electrical insulator. However, boron nitride may be blended with electrically conducting materials, such as titanium nitride or titanium diboride, in order to modify high-temperature electrical conductance and make self resistance-heated small boats and crucibles for the evaporation of aluminium. Billets of boron nitride doped with calcium borate and hot pressed are readily machinable into metal melting crucibles for use at temperatures up to 1500°C under vacuum or 1800°C under nitrogen (Brook et al. 1991). The machining properties of boron nitride, like those of graphite, are considered excellent.

Boron nitride is usable to 1250°C in air, provided the surface film of B_2O_3 is not degraded by contact with other oxides. Boron nitride has only moderate strength because of its low and anisotropic moduli of elasticity. Incidentally, its flexibility is helpful in resisting thermal stress and shock. It has quite good thermal conductivity, which does not change much with temperature and is much less affected by temperature than many other ceramics. Crucibles of pyrolytic boron nitride made by chemical vapour deposition have been useful for high-temperature processing of semiconductor materials. High pressures have been used to prepare cubic BN with higher density and strength (Bullock et al. 1989).

Boron nitride, like graphite, undergoes crystal structure orientation during hot pressing. When isostatically pressed, the solid dense material is anisotropic. Boron nitride is inert and not wetted by a wide range of molten metals. It has low thermal expansion, high thermal conductivity and excellent thermal shock resistance. Its maximum operating temperature is 1600°C depending on grade (Wilson 2000).

3.17 SILICON NITRIDE AND SIALON

Silicon nitride (Si_3N_4) ceramic is stable only under conditions of moderate partial pressure of nitrogen and very low partial pressure of oxygen. It may have applications as a crucible (Richerson and Lee 2018).

Silicon nitride does not occur naturally. It has been synthesized. Some of the powder available commercially has been made by the reaction of silicon powder with nitrogen at temperatures in the range of 1250–1400°C and consists of a mixture of α-Si_3N_4 and β-Si_3N_4 polymorphs. The resulting powder is not of high purity but contains impurities such as Fe, Ca and Al, which were originally present in silicon, plus impurities picked up during crushing and grinding. Higher-purity Si_3N_4 powder has been made by reduction of SiO_2 with carbon in a nitrogen environment and reaction of $SiCl_4$ or silanes with ammonia. Both these methods produce powder of very fine particle size.

The ceramic form of silicon nitride was first produced in 1955 (Xiao and Ralph 2009). This type of material was formed by nitriding silicon powder compacts and came to be known as reaction-bonded silicon nitride (RBSN). During nitriding, the silicon nitride grows within the pre-existing pores of the powder compact, and the original dimensions of the silicon compact remain almost unchanged (linear shrinkage <0.1%) at the end of nitriding. A complex shape can be machined to final size either before nitriding or after partial nitriding.

Silicon nitride has high strength and high hardness due to strong covalent bonds. These bonds also make sintering difficult. In 1961, Deeley et al. achieved high densities by hot pressing previously formed silicon nitride powder using sintering additives. Using magnesium oxide, full density was reached by hot pressing at 1850°C under 23 MPa. The strength of this material was substantially better than that of RBSN.

Sialon ceramics are solid solutions of AlN and Al_2O_3in β-Si_3N_4. They possess excellent mechanical properties, good corrosion and thermal shock resistance, and low wettability by aluminium alloy melt. This material may be a potential candidate for long-term liquid aluminium containment. Sialon ceramics also show excellent corrosion resistance to molten steel and slags.

3.18   ALUMINIUM NITRIDE

Aluminium nitride is a white to pale yellow solid nitride of aluminium, stable to high temperatures in inert atmospheres, and melts at 2500°C. Very pure (100%) aluminium nitride crystals are perfectly water white in colour. In a vacuum, AlN decomposes at~1800°C. In air, surface oxidation begins above 700°C, and bulk oxidation becomes significant above 1370°C. The material hydrolyses slowly in water. Aluminium nitride resists attack by most molten salts, including chlorides and cryolite. The compound was first synthesized in 1877 and can be made by carbothermic reduction under nitrogen or ammonia or by direct nitridation of aluminium. It is difficult to prepare in fully dense form, and hot pressing is used with Y_2O_3 or CaO, as sintering aids, to produce a material sufficiently dense for use. At a high nitrogen pressure, AlN melts without decomposition. This property has been used to produce AlN crucibles without binders, plasticizers or other such additives. The process of ceramic production based on such sintering with partial melting yields AlN crucibles with a high effective density and low porosity (less than 5%). Long and Foster (1959)have given a detailed account of aluminium nitride preparation, fabrication and use.

A unique material with low electrical but good thermal conductivity, AlN possesses, among the ceramics, a rare combination of chemical inertness along with properties, such as thermal shock, linear expansion and erosion resistance, that compare favourably with other materials. It can be used very well as a crucible material. Aluminium nitride is inert to aluminium, copper, tin and calcium melts and does not introduce foreign metal impurities into the notoriously reactive TiAl-based melts either. It is an underexplored crucible material for melting reactive metals (Long and Foster 1959).

High-density AlN crucibles are inert to titanium alloys, yielding non-contaminated castings with good ingot release after solidification of the alloys inside the crucible (Kartavykh et al. 2008, 2009). They are expensive, and large crucible fabrication is difficult. Fabrication of AlN into usable small-size refractory ware by hydrostatic pressing can be done with commercially available equipment. These are best used as coatings on shock-resistant crucibles like Al_2O_3 or graphite. Slurry-based coating methods like air spraying or painting cannot be used for AlN due to its reactivity with water. Plasma spray coating is useful.

Long and Foster (1959) prepared aluminium nitride of highest purity by the process of striking a dc arc between two high-purity aluminium electrodes in a nitrogen atmosphere. The aluminium vaporized in the arc, and the vapour reacted with the nitrogen to form a hard lump between the two electrodes. The nitride lumps, crushed to –100 mesh size, were heated in a graphite crucible in 1 atmosphere argon pressure at 2000–2050°C for 1.5 hours. Ball-milled aluminium nitride compacted by CIP yields green shapes with good handling strength. They are also machinable. These shapes, on sintering in an induction-heated furnace in a graphite crucible, yield products with excellent strength and low porosity even in 1 to 2 hours under argon at 1950–2050°C.

Long and Foster (1959) heated high-purity aluminium metal to 2000°C in aluminium nitride crucibles in both argon and carbon monoxide atmospheres without any noticeable attack on the refractory. Gallium wets the aluminium nitride, as it does most other surfaces. Metallic gallium contained in an aluminium nitride crucible was heated in vacuum for half an hour at 1300°C with no measurable pick-up of metallic contaminants by the gallium.

3.19   COATINGS ON CRUCIBLES

There are rather few instances of a chemically inert crucible also having good thermal shock resistance. In practice, the issue of getting a material with good chemical inertness and high thermal shock resistance is circumvented by selecting a crucible of shock-resistant material and coating it with a layer/layers of inert material. For example, an Al_2O_3 crucible coated with Y_2O_3 (Zhang et al. 2012, Zhang et al. 2013a) results in a crucible with good thermal shock resistance, reasonable chemical inertness to the molten titanium and low cost.

As well as being inert, the coating must be adherent and resistant to abrasion and shock. Y_2O_3 is probably the most investigated coating material for induction melting crucibles (Barbosa et al. 2003, 2005, 2007; Gomes et al. 2011, 2013; Kartavykh et al. 2009; Kuang et al. 2000). It can be applied by the simple process of slurry painting or more sophisticated plasma spraying. The issues in coating with yttria include poor shelf life of the slurry, poor sinterability, low adhesion to many substrates, and if the coating is thick, cracking due to thermal shock. Additives, e.g. SiO_2, are used to improve slurry life and sinterability of the coating, but such low-melting and less chemically inert additives compromise the main property, the chemical inertness, of the coating. For titanium melting crucibles, the thickness of yttria coating is ~200–300 μm.

Porosity is a key parameter for both the crucible body and the coating (Camel et al. 2017; Eastman et al. 1951; Gomes et al. 2013; Tetsui et al. 2010, 2012a). High porosity improves shock resistance and facilitates adhesion of the coating to the substrate. It also leads to higher surface wettability and melt infiltration. Functionally graded coating, where the average particle size varies along the coating depth, has been suggested as a means of fixing the issues of inertness and thermal shock resistance of the coating (Gomes et al. 2011).

Any refractory that is good enough for melting crucibles generally qualifies for making moulds. The melt–refractory reactions are less severe in moulds compared with melting crucibles. There are, however, other issues that need to be addressed.

3.20 SUMMARY

In the beginning, a crucible was used not as a separate entity but as part of the scooped-out earth mended with layers of clay or charcoal dust. The materials used for pottery doubled as crucibles for metallic materials, and eventually, they were made portable. The traditional ceramics, dominated by various clay minerals, and modern ceramics, which included a range of materials from aluminosilicates to many oxide and non-oxide compositions, came into use one by one, very gradually till the 18th century, faster thereafter and rapidly during the past seven decades. These materials became increasingly sophisticated in composition, purity and structure. They possessed superior properties, a wider usage spectrum and higher reliability. Carbon also entered the picture as a crucible material a few hundred years ago and remained unrivalled in superiority for usage in non-oxidizing environments. Carbon, in its many forms and most notably as graphite, is by itself a complete crucible material. Besides, as part of other crucible material compositions, it confers a significant enhancement in properties. Yet, the development of crucible materials, and of crucibles, remains, to date, a work in progress. The important types of crucible materials that are currently relevant for immediate and future uses have been covered in this chapter. The term *crucible* refers to the material entity in contact with the melts being processed– both metallic and slag. As the need for attaining and processing at higher temperatures developed, the term *refractory* came into use to denote high-temperature capability. Refractory ceramics are invariably used as materials for crucibles in metals processing.

Refractory materials for crucibles are broadly grouped into siliceous, aluminous, aluminosilicates, magnesite, dolomite, chrome ore and a miscellany of materials known as special refractories. These refractories are used in two basic forms: pre-shaped objects and unformed compositions or monolithics. Each of these refractory types and forms has its niche application areas. Every crucible material is subject to attack by the melt it is holding. Melt penetration into the crucible and reaction of the melt with the refractory both occur. Generally, refractories have the ability to withstand alteration by penetration, contamination and/or reaction in service and still function as reliable engineering structures. Many factors control this interaction, and some of these are manageable to a certain extent.

Silica (SiO_2) is a major ingredient in refractories, and silica refractories account for about one-sixth of total refractory production. Initially, naturally occurring fireclays and later on, aluminosilicates of a range of compositions and purity levels fulfilled the refractory needs of many major metal processing facilities. The chemical nature of aluminosilicate refractories is acidic, and they are unusable with basic slags and in general environments that tend to be basic.

Mullite as a refractory is highly usable up to about 1600°C. High-alumina bricks have found extensive use as a blast furnace refractory. Alumina is among the most important refractory ceramics and ranks alongside zirconia and yttria in qualities and usability range. Many types of chemically and thermally processed alumina are used in refractories. Tabular alumina has excellent high-temperature properties and thermal shock resistance.

The magnesia or magnesia–lime group includes all refractories made from dolomite and synthetic magnesites. These are the most dominant group of refractories for the basic steel-making processes. Magnesia is noted for its compatibility with iron oxides. Calcia is more reactive with iron oxide. It is also more prone to hydration and disintegration.

For magnesia (MgO), the maximum operating temperature is 1800°C depending on grade. The coefficient of thermal expansion of pure MgO refractory is very high. Hence, the thermal shock resistance is poor. Other refractory raw materials may be incorporated with magnesia in a refractory body to correct this.

The naturally occurring double carbonate dolomite ($CaCO_3 \cdot MgCO_3$) can be converted to refractory dolomite ($CaO \cdot MgO$) by firing. Dolomite bricks are made of sintered dolomite and fused dolomite. To improve the corrosion and hydration resistance, the fired bricks are impregnated with tar or pitch. Currently, the tar/pitch binder is being replaced with a thermosetting phenol–formaldehyde resin. A chemically bonded dolomite brick with 3–4% residual carbon is less wetted by slag and has better thermal shock resistance.

The refractories predominantly used in copper smelting, converting and refining furnaces belong to the mag–chrome composition group. The properties of refractories made from magnesia and chrome ore combination are considered excellent, with one constituent tending to minimize the major weaknesses of the other The properties of the bricks are a function of the magnesite-to-chrome ratio and also the type of bonding. Refractories based on mixtures of MgO and Al_2O_3 have been considered for replacement of mag–chrome due to perceived environmental threats of Cr^{6+}.

The introduction of carbon into magnesia refractories by tar-bonding, pitch impregnation and graphite additions results in much improved performance of the refractory in terms of higher thermal conductivity, good thermal shock resistance and resistance to slag penetration. The C/MgO ratio can be a handle to tailor properties.

Carbon is the only elemental refractory that is also a non-metal. It is unmeltable and is not wetted by metal melts and slag. It dissolves in pig iron melt and to a lesser but significant extent, in certain reactive and refractory metals. It has high thermal conductivity and low thermal expansion coefficient, and this combination imparts to it high thermal shock resistance. The major limitation of graphite is its susceptibility to attack by oxygen. Carbon is used in many forms, mainly as graphite, which became widely available after Acheson developed a process to synthesize it in 1896. In metallurgical processing, the principal uses of graphite are as large carbon blocks in the bottom and bosh of a blast furnace and as electrode cum lining for aluminium electrolysis cells. As an additive, carbon confers an important enhancement in properties on a variety of oxide refractories.

Chemically, BeO is one of the most stable materials. Potentially, beryllia crucibles are excellent containers for liquid metals. Thermal conductivity is relatively high, and for a material of good strength, this gives better thermal shock resistance than Al_2O_3. But, beryllia is toxic. Zircon has excellent refractoriness, good wear resistance, less wettability by molten metal and slag, and hence, good corrosion resistance. Addition of alumina to form a zircon–mullite composition leads to further enhancement in properties.

Zirconia has a combination of high melting point (2715°C) and the potential for high strength, superior resistance to corrosion and erosion to reign among the most important refractories. Zirconia is polymorphic. Around 1170°C, it undergoes the tetragonal/monoclinic transformation, which is associated with a large volume change (~4%). This is the reason why pure zirconia wares break up as their temperature changes through this value. To counteract the deleterious effect of phase changes, zirconia may be stabilized to the cubic phase with small amounts of calcia, magnesia or yttria, with the result that stability, thermal shock resistance and hot load properties are enhanced in the final product. It is also amenable to partial stabilization; cubic materials contain fine dispersions of tetragonal phase. This flexibility, combined with the capability of microstructural control through heat treatment, is the reason for the popularity of zirconia, as it can be engineered to fit a wide variety of special refractory applications.

Zirconia is used in the natural, stabilized or fused state, the last often as a mixture with alumina, silica or other compatible oxides. Zircon with alumina can be processed to form zirconia–mullite grains. This material has high thermal shock resistance as well as good corrosion resistance. With the addition of carbon, the alumina–zirconia–carbon material is used for slide gate plates in steel ladles.

The rare earth oxide yttria is one of the most stable refractory ceramic oxides. Its availability, in high purity, improved with the expansion of the rare earths industry, beginning in the 1970s. Yttria has outstanding chemical stability, but its thermal shock resistance is not very good. This is partly circumvented by fabricating the crucible with a somewhat porous structure. The material is expensive. This led to exploration of its use as a coating on other oxides, and the effort has been rather successful.

Magnesia–alumina spinel ($MgO \cdot Al_2O_3$) and magnesia–chrome spinel are widely used as refractories because of their high melting point of 2135°C, low thermal expansion, good thermal conductivity and high thermal shock resistance. All types of spinel tend to form substitutional solid solutions with each other and with other oxides. In $MgO \cdot Al_2O_3$ spinel, both magnesium and aluminium cations can be replaced by other cations of similar size. This results in the formation of alumina-enriched spinel as well as magnesia-enriched spinel. Alumina-enriched spinel can absorb FeO and consequently, retard the infiltration of slag into spinel-containing refractory. As an alternative to other carbon-free refractories, alumina-spinel fired bricks show superior performance in highly corrosive metallurgical environments. The spinel-containing alumina bricks do not contain carbon, and hence, dissolution of carbon in ultra-low-carbon steel is eliminated. This type of brick is widely used in producing clean steel.

Carbon-containing refractories are used widely in the steel industry. Benefits of carbon include its non-wettability by slag and high thermal conductivity. While carbon confers many of the desired properties on oxide–C refractories, it is also prone to oxidation, and if that happens, the benefits carbon provides will be lost. To overcome this problem, two main types of additive used in oxide–C refractories are metals/alloys (such as Al, Si, Mg and Al/Mg) and boron-based compounds (such as B, B_4C, CaB_6, ZrB_2, $Mg_3B_2O_6$ and SiB_6).

Refractories undergo corrosion when they come into contact with a molten slag or a metallic melt. But, the attack is managed in many ways. A novel new use of slag is as an in situ refractory. This is generally similar to the formation and functioning of a frozen skull in water-cooled crucibles, common in the handling of very reactive metals. In situ refractory is routinely created in the slag splashing operations in BOS vessels. It is also possible to arrange reaction of refractories with the melt contents, generating a protective interlayer between the refractory and the melt.

The composition of the liquid at the point of penetration into the refractory is critical to the corrosion mechanism. A different local composition may arise for several reasons: due to the components of the melt penetrating the refractory at different rates, due to refractories selectively reacting with or dissolving certain melt components, and so on. The phases in the refractory may dissolve in the local liquid, altering its composition compared with bulk liquid. The important consequence of these processes is the resulting significant change in the viscosity of the local liquid, which directly affects its ability to penetrate the refractory and degrade it.

For the use of a ceramic as a crucible material, it is essential that the compound shows no crystallographic phase transitions in the temperature region of interest. This is, in fact, the case for $BaZrO_3$. Apart from a a melting point of about 2600°C, $BaZrO_3$ has a smaller thermal expansion coefficient than ZrO_2 and therefore, better thermal shock resistance. In a binary ceramic, any deviation from the stoichiometric composition will lead to precipitation of the off-stoichiometric surplus at the grain boundaries of the ceramic. This region will be prone to corrosion. This is remedied by additives such as calcia and yttria, resulting in CaO-doped $BaZrO_3$ and yttria-doped $BaZrO_3$. Calcium zirconate ($CaZrO_3$) has a high melting point of 2368°C and chemical stability as good as that of CaO. The zirconate, unlike CaO, is inert to hydration. As in barium zirconate, the formation of off-stoichiometric compounds was restored by incorporating CaO as additive. The crucible was used to melt NiTi and Ti6Al4V alloys at 1500 and 1750°C, respectively, without contamination.

Special refractories include many materials. Silicon carbide, SiC, has good thermal and electrical conductivity, good thermal shock resistance and strength at high temperatures. Silicon carbide bricks find increasing use in the blast furnace lining. They resist corrosion by acid slags, alkali vapour, zinc vapour and oxidizing gases. Silicon carbide is also used in the side walls of Hall–Héroult cells on account of its properties. The first silicon carbide refractories were bonded with clay, and the refractory properties of the bond limited the usability of the material. Self-bonded silicon carbide has removed this limitation. The bonding phase can be SiC itself, silicon nitride or sialon. Silicon carbide crucibles are used extensively for gold and silver melting. They have excellent thermal shock resistance and can be used up to 1500°C. Diborides, especially TiB_2 and ZrB_2, are chemically stable in the environment of a Hall–Héroult cell for aluminium production. They are not wetted by molten aluminium, and their electrical conductivity is high. In aluminium electrolysis cells, replacing the conventional graphite cathode with the diborides has often been seriously considered. Diborides can be used as crucibles for molten metals such as aluminium, copper, silver, gold, zinc, cadmium, magnesium, germanium, lead, bismuth and chromium.

Boron nitride in its hexagonal, soft modification finds use up to very high (~2000°C) temperatures. Billets of boron nitride doped with calcium borate and hot pressed are readily machinable into metal melting crucibles for use at temperatures up to 1500°C under vacuum or 1800°C under nitrogen. Sialon ceramics are based on the solid solution of AlN and Al_2O_3 in β-Si_3N_4. These materials possess excellent mechanical properties at high temperatures, good corrosion and thermal shock resistance, and low wettability by aluminium alloy melt. This material may be a candidate for long-term liquid aluminium contact. Sialon ceramics also show excellent corrosion resistance to molten steel and slags.

A unique material with low electrical but good thermal conductivity, AlN possesses, among the ceramics, the rare combination of chemical inertness and good shock resistance. It can be used very well as a crucible material. Aluminium nitride is inert to aluminium, copper, tin and calcium melts and does not introduce foreign metal impurities into TiAl-based melts either. It is an underexplored crucible material useful for containing reactive melts such as molten aluminium and gallium arsenide. High-density AlN crucibles are not only inert to titanium alloys but also have good shock and erosion resistance. AlN performed well, yielding non-contaminated castings with good ingot release after solidification of titanium melts inside the crucible. These materials are expensive, and large crucible fabrication is difficult. They are best used as coatings on shock-resistant crucibles like Al_2O_3 or graphite.

There are rather few instances of a chemically inert crucible also having good thermal shock resistance. In practice, the issue of getting a material with good chemical inertness and high thermal shock resistance is circumvented by selecting a crucible of shock-resistant material and coating it with a layer/layers of inert material. For example, an Al_2O_3 crucible coated with Y_2O_3 results in a crucible with good thermal shock resistance, reasonable chemical inertness to the molten titanium and low cost. Only a beginning has been made on the use of coatings on crucibles for circumventing metal–crucible interactions. There is a very long way to go.

4 Major Melt—Crucible Systems

4.1 INTRODUCTION

Throughout the history of metals processing, metal makers have faced the challenge of finding and using suitable crucibles for smelting as well as post-reduction melting and casting operations. The challenges have been faced and circumvented with ingenuity, directed experimentation and sometimes plain luck, and there is now available a large and valuable body of knowledge on what works and sometimes, also the underlying reasons for the success. In this chapter, the situation as regards some of the major metal–crucible systems are dealt with not only to look back at the road travelled but also to prepare for the road that lies ahead.

Originally, this chapter was titled 'Metal–Crucible Combinations That Work'. The intention was to consolidate in one place all the different metal–crucible combinations that have been tested and found to work and in the process, to look for the common underlying reasons that make such compatibility a reality. It has become clear in the matter presented so far that materials interact or fail to interact because of certain well-defined reasons. The reasons concern thermodynamics, on the one hand, and transport processes at the interface, on the other. How these factors play out in each individual case is unique, but as is generally explained, there are methods in the madness. These methods are rather implicit and are best revealed by describing the working combination. It was therefore considered best to select certain major metal–crucible systems and look at the key processes involving the crucibles from the point of view of the crucibles.

No metal has relied so much on the availability of suitable crucibles for successful processing as iron and steel have. In fact, the bulk of all the refractories for metal making belong to the sub-group of iron-making refractories or steel-making refractories. Not only has the metallurgy of iron and steel benefited immensely from the development and availability of refractories, but also, the technology of refractories was greatly enriched by the research and development focused on iron and steel. Iron making in blast furnaces; primary steel making in Bessemer converters, open hearth furnaces, Linz–Donawitz vessels and electric arc furnaces; secondary steel making in ladles and degassers; and finally, tundishes, shrouds, slide gates and submerged entry nozzles in the casting system have all relied heavily on the developments in refractories.

It must have appeared that aluminium, with the relatively low melting point not only of the metal itself but also of many of its commercial alloys, would go easy on refractories in processing. This was not to be the case. Whether for primary

DOI: 10.1201/9780429345562-4

aluminium production in Hall–Héroult cells or post-reduction treatments in anode and refining furnaces, the containment problems were serious. Some of the solutions were unique, and because of the temperatures involved, extensive efforts were directed to decreasing the wetting of the refractory by the melt using anti-wetting additives. For other issues, more general melt–refractory interaction management techniques were used.

Copper is not particularly known for its reactivity, but Mark Schlesinger noted that furnaces used for producing molten copper face exemplary challenges to the refractory life in the form of highly aggressive slags, mechanical stresses, batch operation and higher operating temperatures. Refractories from the mag–chrome family fulfil nearly all the refractory needs in copper processing. However, two drivers are forcing a change. One is the objection from environmental authorities regarding chromium, and the other is the use of higher process temperatures, off gas richer in SO_2 and the use of calcium ferrite slag.

Beginning around 1940, activity started picking up in developing processes for the production of many metals that had till then seen little life outside laboratories. They are what eventually came to be known as less common metals and included metals essential for emerging new technologies like nuclear energy, aerospace and a little later, the electronics, telecommunication and biomedical industries. At that point, the little-used technique of metal replacement reactions became an important process for the 'smelting'. In a popular version of this technique, the reduced metal is produced in a molten form before allowing it to solidify as an ingot. Thermal and reactivity management are key factors, and both were enabled by the use of a suitable refractory lining of the reduction reactor, known popularly as the 'bomb reactor'. In an equally versatile format, the metallothermic reductions were also carried out to yield the metal as sponge. Not that there was much choice of obtaining these metals in any other form, because handling metals like titanium, zirconium or hafnium in molten form at the reduction stage would have presented formidable crucible problems. The flagship process for obtaining reactive metals in sponge form is the Kroll process, which is based on controlled reaction between purified metal chloride and molten magnesium in a protected atmosphere at temperatures in the range of 950°C. As long as sharp temperature excursions are avoided, steel and Inconel reactors are fine for crucibles and have been widely used. While the Kroll-type process can be used for the preparation of some of the rare earth metals, a more robust method, relying on calcium reduction of rare earth trifluorides, was eventually used for the preparation of rare earth metals. The process was carried out in induction-heated tantalum crucibles in a protected atmosphere. To minimize ingestion of tantalum into the metal, a relatively low-temperature reduction based on lithium vapour as reducing agent was also used. Tantalum, molybdenum and titanium wares were used as contact materials here. A Pidgeon-type process based on reduction–distillation was also used for certain rare earth metals. As well as all these, electrolytic methods were developed for most of the rare earth metals, using graphite, molybdenum and tungsten as construction materials. The high point is the production of yttrium in molten form in an electrolysis cell using fluoride electrolytes operated at 1700°C.

Uranium is a metal that had to be processed frequently for a variety of purposes in a molten form. There were also requirements to evaporate it from a crucible. All this meant strong focus on the development of uranium melt-compatible crucibles. Vanadium is well established as an alloy additive to steel, and not much was known of its use as a metal or an alloy base. In fact, for three decades, there was just one monograph on vanadium, written by William Rostoker in 1958, until another one appeared in 1992. This was until interest in its possible role as a construction material for liquid metal loops in fast (nuclear) reactors and thermonuclear fusion reactors brought it back into focus. Vanadium is notoriously reactive and impossible to melt for any length of time without contamination in a metal or ceramic crucible.

Somewhat less aggressive than vanadium, but still not easy on crucibles and moulds, is titanium. It is exceedingly important, both technologically and financially, to find a crucible for routine melting and casting of titanium and its alloys. The importance and urgency stem from its relevance as a premium aerospace, energy conversion and biomedical material base, and in many cases, it is not replaceable in its applications without technology and cost penalties. After nearly seven decades of directed efforts, the interesting situation is that it may still be premature to assert that we have succeeded.

What follows in this chapter is a crucible-centric exposition of the relevant processes for all these metals. The objective is to relate what is known about metal–crucible interactions at the basic level to the interactions playing out in real processes. This will help the task ahead as regards managing new metal–crucible combinations and suitable approaches that will improve the chances of success.

4.2   IRON AND STEEL

Few metals have been as closely dependent on the availability of crucibles for their extraction as iron. Since iron was discovered and recognized as a highly usable metal, for over two millennia, it was not prepared in molten form. All processes to make products from iron were designed to work around the need for melting it. The reports on the first iron castings being made in China also refer to their access to fairly refractory clays available only in that country. Eventually, the development of refractories for iron and steel became an important area of study and industry and led the way for the development of the entire industry of metallurgical ceramics and refractories.

Even though iron is the second most abundant metal in the earths' crust (5.8%) after aluminium (8.3%), iron is the most used metal on the planet, mostly as steel. The iron-rich ores (haematite [Fe_2O_3], magnetite [Fe_3O_4]), are reduced using coke in an iron blast furnace. The reduction process predisposes the as-produced iron, called the hot metal or pig iron, to be liberally contaminated with a variety of impurities derived from the ore, the reducing agent and the various additives and fluxes that were put in to make the process efficient and smooth. There are also the inevitable impurities that the iron picks up from the furnace structure. The impurities include carbon, silicon, phosphorus, aluminium, manganese and sulphur, and many more in minor quantities. Amazingly, the metal is still usable as such for a few applications (after a certain amount of metallurgical tending), but as the impurities are progressively decreased both in number and in concentration, the variety of uses increases greatly and the uses become more sophisticated.

Many elements are incorporated into iron and steel by design, for enhancement of properties. The process metallurgy part of iron and steel is all about impurities control, i.e. the removal of impurities from the metal and protection of the metal from re-ingestion of impurities during further processing to a finished product. It is amazing that steel is the only 'metal' that has the adjective 'clean' as part of its name. 'Clean steel', which has less than 50 ppm of residual impurities (PCSNHO), is the cleanest metal produced routinely under industrial conditions in tonnage quantities (Poirier 2015). The processes and materials used represent some of the finest and most ingenious in process metallurgy. It is the scale of operation that is daunting. Besides, the physical chemistry and metallurgical features of these operations are so varied and so deeply challenging that developments in the materials and techniques that constitute the making and treating of steel eventually became the core of the evolving metallurgy practice and knowledge base. When confronted with challenging technical issues regarding materials and processes, it is not unusual for metallurgists, irrespective of their specialization, to look for analogies in iron and steel metallurgy. One such issue is the interaction of melts with the crucibles or containment and the consequent effect on properties.

Blast furnace iron making is the most popular process for extraction of iron from its ores. Pure iron has a high melting point (1536°C), but when it dissolves carbon, the melting point decreases significantly, and the iron–carbon eutectic occurs at 1150°C and 4.2% C. Steel, the alloy of iron and carbon with C content less than 4.2%, has the ability to take many other elements, such as chromium, nickel and manganese, into its composition, and the presence of these elements modifies and usually enhances one or more properties of the steel. There are hundreds of grades of steel characterized by compositions and microstructures, and they are produced through secondary steel-making processes.

The primary steel-making process largely follows the blast furnace–LD process/basic oxygen furnace (BF–LD/BOF) route and to a lesser extent, the directly reduced iron–electric arc furnace (DRI–EAF) route. Secondary steel making covers the refining of primary steel with respect to composition and cleanliness before casting. Secondary steel making involves deoxidation or killing with Al/Si additives, vacuum degassing, desulphurization, homogenization, alloying, and control of inclusion chemistry and content. These operations are commonly performed using ladle furnaces (Fruehan 1998).

4.3 BLAST FURNACE REFRACTORIES

The functioning of a blast furnace depends on the performance of the lining refractory (Biswas and Sarkar 2020). These refractories undergo exemplary physical, thermal and chemical stresses, and the blast furnace has long been the test track for new and improved refractories. The main sources of physical stresses are gas flow through the stack, erosion by descending burden, and flow of liquid iron in the hearth. The thermal stresses are due to the heat load inside the furnace. High-temperature reactions of refractories with gases in the stack and belly area, corrosion reactions with slag at the belly and bosh, and penetration of hot liquid metal in the hearth, in combination with the physical and thermal stress factors, cause the degradation of refractories. Alkali infiltration and reactions with the bonding phases, especially in the bosh and lower stack, have the greatest effect on the lining. The alkalis come with the coke.

The blast furnace slag contains, as well as other components, FeO (+2, basic oxide) and alkali vapour. These are the components that attack the refractory. The various reactive species and the temperature range in which they are effective are CO (on stack refractory: 450–850°C), alkali (800–950°C), zinc vapour (750–900°C) and O_2 (on C refractory: >400°C),as well as oxidation by CO_2 and H_2O (>700°C). Hence, the lining refractory needs to be kept at temperatures lower than these temperatures of reaction by external cooling. This 'thermal solution' is implemented by cooling plates and staves, cigar coolers and spray coolers. All portions, including the tuyere zones and the hearth, are cooled and the bias has been towards refractories that have good thermal conductivity.

In the conventional refractory lining in the blast furnace stack and bosh, the lining was made of high-alumina bricks containing 62–85% Al_2O_3. Even in the 1960s and 1970s, the refractory linings in blast furnace construction, including the hearth, were made of different qualities of high-alumina bricks. Their performance and life were limited due to poor alkali resistance, poor thermal shock resistance and also very low thermal conductivity. This made the cooling of the lining ineffective. The attack on high-alumina refractory by slag and alkalis resulted in the formation of compounds that melt at <1400°C, resulting in severe corrosion of the refractory during furnace operation.

The use of a carbon hearth in 1970 and of silicon carbide (SiC) refractory in the bosh, belly and lower stack area came as major developments for modern blast furnace lining. Since the turn of the century, there has been no brickwork in the new-generation shaft and belly of blast furnaces in newly built or rebuilt furnaces. Nitride- or oxy-nitride-bonded SiC bricks with coolers embedded in the stave are used in the stack, and micropore carbon or graphite blocks are used in the belly and bosh areas. For the hearth and bottom of blastfurnaces, semi-graphitized carbon blocks and microporous carbon blocks are used. The carbon blocks used are all of high heat conductivity. Superior-quality carbon blocks of micropore and super-micropore quality (pore size <1 μm) and very high thermal conductivity are used.

Carbon hearth refractories are in direct contact with liquid iron. Under a high ferro static load, the molten metal penetrates through the pores of carbon blocks and embrittles the penetrated layer, which eventually dislodges due to turbulence in the melt. This causes progressive thinning of carbon blocks. Penetration of hot metal and eventual thinning may be avoided or minimized if the radius of open pores of carbon blocks is small enough.

4.4   STEEL-MAKING REFRACTORIES

As mentioned in Chapter 1, the original Bessemer converter had an acid (siliceous) lining. Its usage was restricted to a very low phosphorus- and sulphur-bearing pig iron, since these elements are not removed in the process. In the Thomas process, also referred to as the basic Bessemer process, the vessel was lined with basic refractory (burned limestone) to handle the slag with a high basicity. Calcia was added as flux during the air blow to form a basic slag with high CaO content, which was able to remove phosphorus as calcium phosphate.

The hearth of Siemens' furnace (open hearth furnace) was constructed of acid bricks topped by a layer of sand. Later, the siliceous bottom of the acid furnace was changed to permit the charging of limestone and use of a basic slag for the removal of phosphorus. The hearth was constructed with a lining of magnesite brick, covered with a layer of burned dolomite or magnesite.

A schematic cross section of an oxygen steel-making vessel is shown in Figure 4.1. The refractories used in BOF linings are pitch-bonded magnesia or pitch-bonded dolomite, which are made with resin bonds, metallics, graphites, and sintered and/or fused 99% pure magnesia (Fruehan 1998).

All basic oxygen steel-making processes, top, bottom and combined blowing, rely on oxidation by externally supplied oxygen gas. Oxygen is blown into the melt, and dissolved elements, such as C, Si, P and some S, are oxidized to form CO, CO_2, SiO_2 and P_2O_5. Manganese and iron are oxidized to MnO and FeO. The CO and CO_2 gases escape out, and other oxides collect to form an acidic slag. All these slag-forming oxides can react with basic refractories and corrode them. CaO (burned lime), MgO (burned dolomitic lime) and CaF_2 (fluorspar) are added to the molten bath to neutralize the acidic nature of slag, and form a basic slag, in the later part of the blowing. The added lime is fluxed and dissolved by the siliceous slag, and the resultant basic slag absorbs sulphur and phosphorus from the molten steel and also protects the lining from attack by FeO and SiO_2 (Biswas and Sarkar 2020). MgO may be introduced as dolomitic lime to saturate the slag or even to supersaturate with MgO. Supersaturation results in a MgO build-up on the vessel walls, which serves to extend the life of the lining.

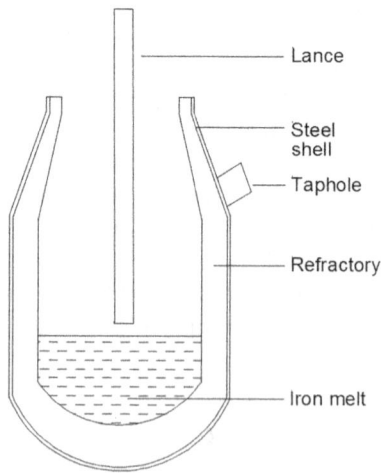

FIGURE 4.1   Schematic of a Linz–Donowitz vessel (From Fruehan, R.J. (Ed.). 1998).

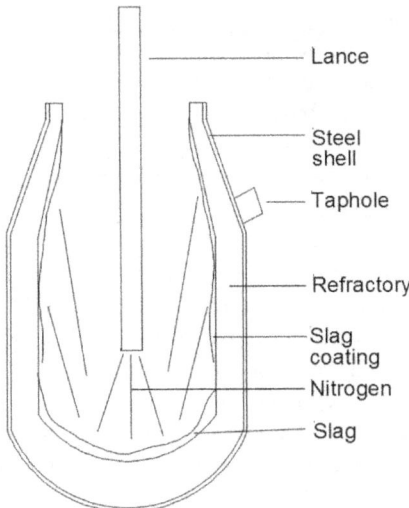

Lance

Steel shell

Taphole

Refractory

Slag coating

Nitrogen

Slag

FIGURE 4.2   Schematic of a Linz–Donawitz vessel – slag coated (From Fruehan, R.J. 1988).

The technology known as slag splashing evolved in the early 1990s. After tapping a heat, it entails the use of the oxygen lance to blow nitrogen on the residual slag, splashing it to coat the walls and cone of the converter (Figure 4.2). Slag splashing seeks to utilize the technique of working off a slag coating instead of the working lining brick or its gunned coating. (Fruehan 1998).

Until about 1976, the main refractory used in BOF was dolomite bricks, such as tar-bonded dolomite or high-fired magnesia-enriched dolomite. During the 1980s, the use of MgO–C bricks started due to their better corrosion and thermal shock resistance. MgO–C bricks were used to line the whole vessel, and that practice has continued to the present. As already mentioned, the durability of MgO–C bricks is significantly improved if the dissolved MgO in BOF slag reaches or exceeds the saturation point of MgO in slag at the operating temperature.

Carbon is used in refractories for its excellent thermal shock resistance and corrosion resistance towards BOF slag and steel. But, it has an inherent drawback of susceptibility to oxidation, and it also increases the thermal conductivity of the lining, which results in heat loss and increase in shell temperature. The development of nano-carbon-containing MgO–C bricks with reduced total carbon content has improved oxidation resistance and reduced thermal conductivity. The addition of nano-carbon serves to lower the total graphite content of conventional magnesia carbon bricks without sacrificing the beneficial effects of carbon addition (Biswas and Sarkar 2020).

4.4.1   REFRACTORIES FOR SECONDARY STEEL

Dolomite and dolomite–magnesia bricks are the most commonly used refractories for the Argon Oxygen Decrburization (AOD) vessel. The very basic nature of dolomite refractories permits operation with higher-basicity slags, which enables improved chrome recovery, and desulfurization. Magnesia–chrome and magnesia–carbon bricks are also used in AOD. The same types and grades of refractories are used in Vacuum Oxygen Decarburization (VOD) ladle applications, but with more equal amounts of magnesia–chrome and dolomite–magnesia bricks. The choice of brick composition, magnesia–carbon or fired magnesia–chrome, is based onthe carbon pick-up tolerated during processing.

In an RH (Ruhrshahl Heraeus) degasser, working lining refractories are exposed to a variety of conditions depending on location. The linings are often zoned (Fruehan 1998). All locations are exposed to thermal (changes between 200 to 600°C) and pressure (ambient to 70 Pa) cycling. Apart from liquid steel at 1650°C, snorkel and lower vessel refractories face the erosive action of high-velocity, turbulent steel flow. RH and RH-OB (oxygen below) degasser working linings are usually direct-bonded magnesia–chromite bricks containing fused grain, or 50–60% alumina brick, or 90% alumina, or tabular alumina, or spinel castable. Castables with high-purity materials are used for the snorkel exterior.

4.4.2 Refractories for Casting

In casting, it is desirable to preserve the purity and cleanliness of steel achieved in secondary metallurgy operations. The machines and refractories are chosen accordingly.

In 1940, the lowest level of total impurity (C, S, O, N and H) content possible in steel was a few hundred ppm, and by the turn of the century, the value was less than 100 ppm. It dropped further to less than 30 ppm by 2015 (Poirier 2015). As well as being present in solid solution, oxygen is often present as oxide inclusions of various sizes in steel. They are both endogenous or micro-inclusions (originating from within the steel) and exogenous or macro-inclusions (e.g. eroded and entrapped lining). The possible influences of non-metallic elements on various properties of the metal have been summarized by Poirier (2015) and listed in Table 4.1.

Lower values (ppm)currently attainable for many elements are shown in the last row of the table. For steel produced routinely in industrial conditions, the values are somewhat higher. The overwhelming influence of oxygen on a range of steel properties cannot be ignored. Next comes sulphur.

As emphasized already, purity achieved in secondary metallurgy should not be lost during continuous casting. The refractory materials directly determine both the type and the amount of solute elements such as carbon, oxygen and the non-metallic inclusions. The interactions between metal, slag, atmosphere and refractory products are related to the chemical and mineral composition, microstructure and phase distribution of the refractory. The key reactions are direct dissolution of the refractory with or without precipitation, dissociation, volatilization, redox reactions between an oxide and a metallic element, and combination of the refractory and a non-dissolved element present in the steel (inclusion).

When the steel melt is re-oxidized by the refractory lining, the oxygen reacts with dissolved Al (aluminium) in steel to form Al_2O_3, which partly floats into slag

TABLE 4.1

Influence of Non-Metallic Elements on Steel Properties

Properties	Hydrogen	Carbon	Nitrogen	Oxygen[a]	Phosphorus	Sulphur*
Internal soundness	x			x		
Surface defects				x		
Fatigue				x		x
Bending				x		x
Deep drawing		x				
Anisotropy				x		x
Toughness			x	x	x	x
Weldability			x		x	x
Electromagnetic properties		x				
Lower limits of residual elements in steel making, ppm	<1	5	10	5	10	5

Source: Poirier, J. 2015.

[a] (Control of inclusions)

and partly remains in steel as inclusions. Very fine inclusions (the smallest size) may not really float up for skimming or to be collected by the slag. Refractory materials with the lowest oxygen potential result in lower aluminium losses and hence, a decrease in inclusions by re-oxidation. Lime, corundum, magnesia and high-purity synthetic materials such as alumina spinel, MgO–C and doloma–C are examples of such materials. If present even in small amounts, SiO_2, Cr_2O_3 and ZrO_2, finally result in a high inclusion content.

Shaped refractory parts (Figure 4.3) are used to guide and protect liquid metal from the ladle to the mould where steel is solidified in continuous steel casting (Poirier 2015). Materials of the submerged nozzle have a role in clogging and unclogging, which leads to metal contamination by alumina particles or clusters. The tundish lining can also have a similar role. Contamination can also be caused by the ladle shroud tube, where reactions similar to those in submerged nozzles can occur, and by the sliding gate system, where the plates can undergo degradation.

Submerged nozzles are often alumina–graphite products. Clogging of submerged nozzles by alumina build-up is a major cause of dysfunction in continuous casting of aluminium-killed steel. The mechanism described by Poirier (2015) focuses on the deposition of Al_2O_3 as a result of the thermochemical reduction of nozzle constituents coupled with the oxidation of the aluminium in the steel. Carbon is the key element. Alumina–graphite refractories contain secondary phases and impurities such as SiO_2, Na_2O and K_2O, which can be reduced by the refractory carbon at steel-making temperatures, and generate gaseous species. The carbon monoxide so generated oxidizes aluminium in steel to alumina, which builds up as a solid

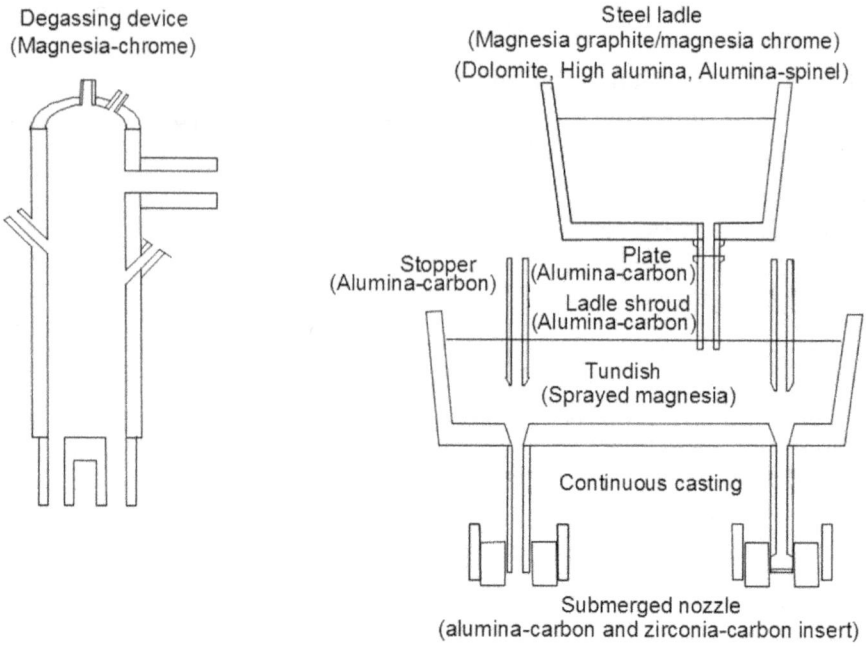

Degassing device
(Magnesia-chrome)

Steel ladle
(Magnesia graphite/magnesia chrome)
(Dolomite, High alumina, Alumina-spinel)

Stopper
(Alumina-carbon)

Plate
(Alumina-carbon)

Ladle shroud
(Alumina-carbon)

Tundish
(Sprayed magnesia)

Continuous casting

Submerged nozzle
(alumina-carbon and zirconia-carbon insert)

FIGURE 4.3 Main refractories in relation to steel quality and metal cleanliness (secondary steel making and casting) (From Poirier, J. 2015).

deposit. Avoiding carbon from the refractory should eliminate some of this alumina deposition. Similar improvements can also be expected using highly pure Al_2O_3–C mixtures with as few silica and alkaline impurities as possible.

The tundish refractory is usually made of magnesia and forsterite ($2MgO \cdot SiO_2$) raw materials. The lining may thus react with Al in steel, more so if the lining is porous. At the steel/refractory interface lining, a layer composed of MgO–Al_2O_3 spinel forms by reaction between aluminium in steel and forsterite in the refractory. Spalling due to the mismatch of properties between the spinel layer and the MgO–forsterite refractory lining can lead to MgO–Al_2O_3 inclusions in the steel. The ladle shroud is usually made of alumina–graphite. The same reactions as in the submerged nozzle are also possible here.

The sliding gate system for steel ladles is used to control metal flow during teeming, and consists of a mechanical assembly containing the refractory plates (Poirier 2015). Air leakage through the worn plates has adverse effects on the cleanliness of the steel and the wear of the refractory by corrosion.

Al_2O_3–C slide plates have been widely used. They are produced using tabular alumina, fused white alumina, and graphite with Si and Al metal powder addition. However, these plates may not be ideal for producing low-carbon or ultra-low-carbon clean steel because of the possible diffusion and dissolution of carbon into liquid steel. The alternatives are SiO_2-free Al_2O_3–C plates and spinel-containing plates such as $MgO \cdot Al_2O_3$–C and MgO–C. Al_2O_3–ZrO_2–C plate is currently used to cast Ca-treated steel.

4.5 THE HALL–HÉROULT CELL

The invention of an electrolytic process for aluminium extraction was outlined in Chapter 1. Almost all the world's aluminium is produced by electrolysis in Hall–Héroult cells. During the electrolysis of alumina dissolved in molten cryolite, aluminium ions discharge at the carbon cathode and collect as molten metal at the prevailing temperature (950–965°C) in the cell. The anions discharge at the carbon anode and evolve as a mixture of carbon oxides: 20–40% CO and 60–80% CO_2. The melts of aluminium and of the electrolyte are non-miscible. In the cell, electrolyte (density = 2.1 g/cm³) remains above the molten aluminium (density = 2.3 g/cm³). Typically, the depth of the molten aluminium layer is about 15–40 cm, and the depth of the electrolyte layer is 15–25 cm (Yurkov 2017).

Pure cryolite, Na_3AlF_6 ($3NaF \cdot AlF_3$), melts at 1010°C. The melting temperature of the electrolyte is brought down to 930–960°C either by adding fluorides (calcium fluoride, magnesium fluoride) or by varying the ratio of sodium fluoride (NaF) to aluminium fluoride (AlF_3). The cell, shown schematically in Figure 4.4, is a shallow bath tank (0.6–0.8 m deep), 10–15 m long and 3–5 m wide. Electrolysis is conducted at 4.16–4.20 V, keeping the space between the anode and cathode at 40–60 cm. The

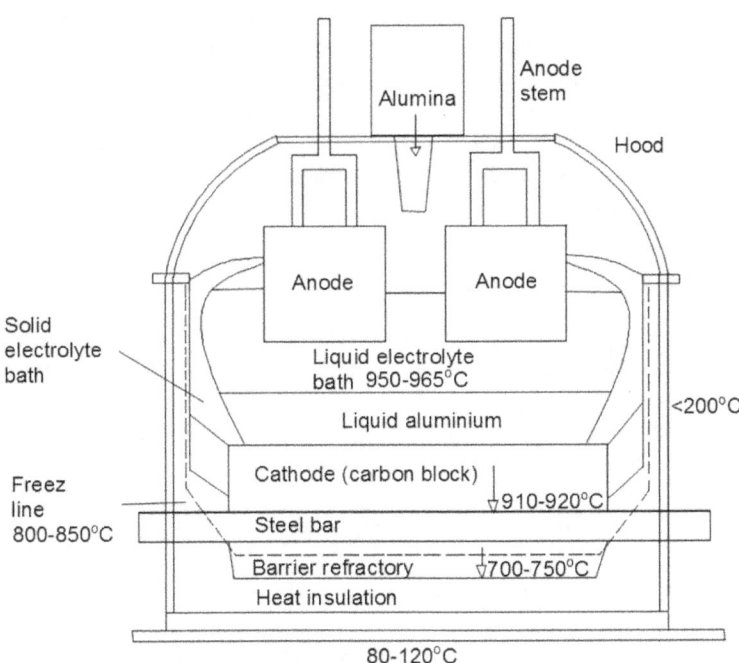

FIGURE 4.4 Hall–Héroult Cell (From Yurkov, A. 2017).

bath is covered by a frozen electrolyte crust. The cells are lined with carbon cathode bottom blocks and a side lining. Below the cathode block are the refractory layer and the heat insulation layer.

The issues of metal–crucible interaction in the Hall–Héroult cell mainly concern the interaction of the molten metal and the bath components with the cathode and the refractory layer. Even though the temperatures involved are relatively low, interactions are many, and intensity is high. Molten aluminium and electrolyte in operating cells are in a state of constant movement, at a speed that may range from 1.5 to 30 cm/s, in the electromagnetic environment prevailing in the cell (Yurkov 2017).

During operation, the electrolysis current also heats the cell. A considerable amount of excess heat is generated in electrolysis (i.e. more than what is required to maintain the bath at 950–965°C), and is dissipated through the side walls of cells. Approximately half of this heat is dissipated through the side wall, and the other half goes away through the bottom and upper parts. Therefore, the side lining, made of SiC blocks or carbon side-wall blocks, has minimal or no heat insulation, while the bottom has good thermal insulation. The temperature profile is also maintained such that a certain amount (a few centimetres thick) of electrolyte freezes on the side lining and protects the lining materials from the molten cell contents. Freezing of electrolyte is not permitted in the bottom within the cell but for a small portion near the side wall. Insulation is therefore differently arranged in this portion. The thermal balance will be disturbed if the electrolyte infiltration through the carbon cathode (block) goes past the refractory layer and ends up in the heat insulation layer.

In the design and construction of the cell, the objective is to have the 800–850°C isotherms within the main body of the upper, protective refractory layer (Siljan et al. 2002). The solidus of the advancing melt falls in this temperature window. The most important purpose of the refractory layer, i.e. as a barrier to bath intrusion and to prevent deterioration of the thermal insulation layer below, will then be served. Even the thermal conductivity of barrier layer refractories, if infiltrated by electrolyte, may change by a factor of 3–5.

4.5.1 Interactions in the Cell

Key interactions to be considered in the cell involve the electrolyte, molten aluminium and sodium with the cell materials, particularly the cathode carbon block and the refractory beneath it.

Sodium is generated and also transported through the cathode carbon due to reactions given here (Siljan et al 2002):

$$\text{In bath:} \quad 2Al + 6NaF = 2AlF_3 + 6Na\left(Al\right) \tag{4.1}$$

$$\text{Metal pool/cathode:} \quad 6Na\left(Al\right) = 6Na\left(C\right) \tag{4.2}$$

There are two main pathways for sodium to go past the cathode and reach the barrier refractory layer of a reduction cell: sodium diffusion (intercalation) in the carbon lattice and capillary flow of electrolyte through permeable pores of carbon cathode blocks and carbon of ramming paste. The resulting sodium pressure at the lower surface of the cathode block might reach as high as 0.02–0.03 atm.

The radius of a sodium atom being significantly larger than the distance between the interlayers of carbon atoms, the intercalation of sodium atoms in the crystal lattice causes swelling and eventually, cracking of carbon blocks. Cryolite (and also aluminium) will penetrate through the cracks to the bottom of the cathode. More than the dry penetration of sodium, it is due to 'wet' penetration, or the capillary flow of electrolyte through permeable pores of carbon cathode blocks and rammed carbon of ramming paste, that the deterioration of the refractory lining occurs.

Thus, as well as sodium and electrolyte, aluminium also ends up at the other side of the cathode due to damage to the cathode. The pathways created by the damage allow liquid aluminium to penetrate down to the top of the refractory lining. Then, a number of reactions occur between the penetrated materials and the refractory, which result in the formation of new phases and cause volume expansion and also viscosity changes in the melt.

4.5.2 BARRIER REFRACTORY

Typically, the barrier refractory layer is made of aluminosilicate (usually fireclay) bricks or unshaped dry barrier mixtures. The aluminosilicates are prone to infiltration and attack by molten electrolyte components. Sodium and molten fluorides cause significant mineralogical transformation in the refractory.

Many chemical reactions are possible among cryolite, sodium fluoride, alumina, silica and mullite. The reaction products are albite ($NaAlSi_3O_8$) and nepheline ($NaAlSiO_4$), which may coexist with cryolite. The volume effect of the reactions is the real cause for concern. Depending on the conditions, the products of reaction between the infiltrated electrolyte and the refractory barrier layer end up in the form of a convex lens. This lens can grow up to almost a metre in thickness and press on the carbon cathode blocks from below, causing bending strains and cracking. The upper part of the lens (immediately below the cathode block) comprises albite, nepheline, cryolite and sodium fluoride. The middle part of the lens consists of albite, nepheline, cryolite, sodium fluoride, quartz and mullite, and the lower part of the lens consists of reacted material and of 30–50% of unreacted fireclay (Siljan et al. 2002).

One remedy suggested to diminish bath penetration through the cathode block to the refractory level is to limit average pore size in carbon cathode blocks to 5–15 μm. Bath penetration is facilitated if the pores in the carbon cathode blocks are above 25–30 μm. With bath components at the doorstep of the refractory layer, Siljan et al. (2002) singled out the silica content of the aluminosilicates as the major handle that can be used to stop the further progress of the reaction. By exposing samples of firebricks to a fluoride melt at 950°C ('cup test') as well as to sodium vapour at 800°C, Siljan et al. (2002) found that materials high in silica deteriorated less than materials rich in alumina. The reason is that the silica-rich materials formed larger amounts of a viscous glass phase. A fluoride melt, which has penetrated through the carbon cathode, accumulates below the cathode carbon, while sodium aluminosilicate, which is denser than the fluoride, will settle below it and above the refractory lining. The continued contact and further reaction of the fluoride melt with the refractory is retarded by the dense viscous layer of the sodium aluminosilicate-based melt. This has been revealed in autopsies of shut-down cells as well as in laboratory-scale cup tests.

The composition of melt that results from the interaction of the infiltrated materials with the aluminosilicate refractory depends on the proportions of silica and sodium fluoride in the reactants (Siljan et al. 2002). The key components are albite and nepheline,which strongly determine melt viscosity, because melt viscosity apparently plays a larger role than pore size of the refractory as regards eventual melt intrusion into the refractory. The solute in the melt that increases the viscosity is albite. The solubility of albite in cryolite is also high. It is good to have albite in the melt. The viscosity of nepheline-rich melt isseveral orders of magnitude lower compared with the viscosity of albite-rich melt. In practice, this (albite-rich melt) is achieved by selecting a refractory with silica content above 60% and by a controlled input of sodium compounds, e.g. with excess sodium fluoride.

There are two choices. To freeze out the fluoride melt at an isotherm located not far below the cathode carbon, i.e. to ensure a low penetration depth into the refractory lining, a high solidus temperature is preferred. This is defined by the refractory material and the penetrating bath. For this, higher alumina is needed in the refractory composition. However, with a substantial amount of silica in the refractory lining, the viscosity of the melt increases so much that penetration is in any case restricted to a low depth. This aspect requires a silica-rich refractory (>53 wt.% SiO_2). The approach favoured by Siljan et al. 2002 is to focus on managing the viscosity of the penetrating melt. They did also point out that the likely gain in solidus by increasing alumina content in the refractory is not so substantial.

For the record, some of the chemically effective refractory barriers against the penetration of cryolite are tin oxide, nickel oxide, compounds of nickel oxide, iron oxide or zinc oxide (such as spinel $NiFe_2O_4$). These oxides almost do not react with NaF and aluminium fluoride. But, their cost is impractically high.

4.5.3 SIDE LINING MATERIALS

Anthracite-based carbon side blocks were used for many years before being replaced with SiC in the 1980s. The advantages of SiC side lining compared with carbon side lining are better corrosion resistance to electrolyte and aluminium, higher oxidation resistance and higher thermal conductivity. Currently, nitride-bonded silicon carbide is the only material used as side lining (Yurkov 2017).

4.6 POST-REDUCTION TREATMENT OF ALUMINIUM

As-reduced aluminium from the Hall–Héroult cell is 99.5 to 99.7% pure. It contains non-metallic (electrolyte, alumina, aluminium carbide, carbon particles and oxides), metallic (iron, silicon, titanium, sodium and calcium), and gaseous (hydrogen) impurities. These impurities are removed in post-reduction treatments by fluxing (and skimming) and degassing before casting the 99.8 to 99.85% Al into moulds.

The metallurgical equipment of cast houses at smelters includes ladles for Al transportation, stationary or tilting holding furnaces, degassing units, induction furnaces and gas-fired furnaces. There is almost no difference in the constructions of melting and holding furnaces. Usually, melting furnaces are heated by natural gas, and holding furnaces are heated by natural gas or by electric heaters. The general requirements for refractories in these facilities are porosity <15%, average pore size <5 μm, and low wetting by Al.

4.6.1 REFRACTORIES IN ALUMINIUM MELTING AND HOLDING FURNACES

Schlesinger(2006)in his book on aluminium recycling observed that for its relatively low melting point, aluminium places a rather disproportionate demand on materials usable for its containment. He then listed the nine requirements that the refractories in aluminium melting and holding furnaces need to fulfil: (i) lack of reactivity with the molten aluminium, (ii) low solubility in molten flux, (iii) lack of reactivity with the Al_2O_3 in the dross, (iv) low thermal conductivity, (v) hydration resistance, (vi) oxidation resistance, (vii) stability at temperatures up to 1200°C, (viii) good mechanical strength and (ix) low cost.

As is common in such situations, no single refractory meets all the requirements. The job is done by a combination of refractories. As an example, in the wall construction in a reverberatory melting furnace for aluminium, a corrosion-resistant refractory is used at the hot face and a low-density insulating material between it and the steel shell. The purpose is to locate the 'freeze plane', below which penetrating metal solidifies, well within (Dunsing 2002) the thickness of the corrosion-resistant refractory lining. This is a rather universal requirement implemented in melt containment configurations. It was seen in the Hall–Héroult cell also. The insulating brick is not sufficiently resistant to molten metal and is also porous, and will not survive any contact with the melt. If the molten metal penetrates and ends up beyond the thickness of the refractory facing, and freezes between the refractory and insulating material, not only is the insulation arrangement ruined but the materials also become unusable.

Such a combined refractory structure (Figure 4.5) works in the molten aluminium transportation ladle (Bonadia et al. 2006) as it does in the torpedo ladles of molten iron transportation. This arrangement, in principle, works in crucible induction furnaces also – a multilayer refractory lining comprising a chemically stable working lining to contact the melt and a thermal insulation lining between the working lining and the vessel walls (Figure 4.6). Phosphate-bonded high-alumina bricks resist aluminium attack well due to their unique microstructure, i.e. small pores, which significantly decreases metal penetration. This is a typical working lining. There is a thin layer of highly insulating microporous silica between the steel shell and the intermediate brick layer, on the wall and at the bottom, for thermal insulation. A semi-insulating castable coupled with ceramic–fibre insulation is used for the ladle lid.

The crucible induction furnace shown schematically (Figure 4.6) is used for making alloys and master alloys and for remelting scrap (Yurkov 2017). The lining of

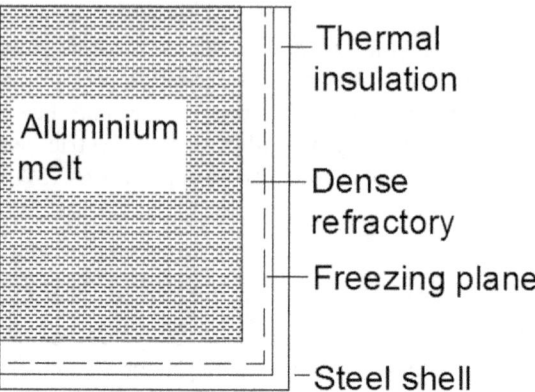

FIGURE 4.5 Composite refractory structure – aluminium transportation ladle (From Schlesinger 2006, Bonadia et al. 2006, Yurkov 2017).

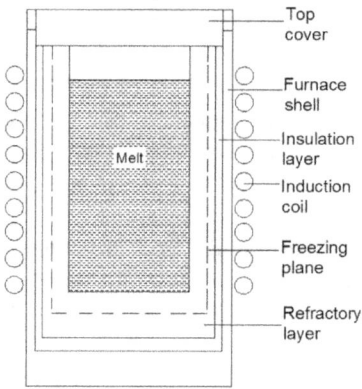

FIGURE 4.6 Composite refractory structure – induction furnace (From Schlesinger 2006, Yurkov 2017).

induction furnaces is made from ramming mixes as well as plastic mixes. An anti-wetting additive is normally used. Anti-wetting additives decompose, leaving a bulky residue. A fine pore -size structure with a small pore dimension is a more reliable option in such conditions. Phosphate-bonded bauxite bricks or low cement castable (LCC) with filler from bauxite are used. Degassing units are used for the refining of Al alloys from gases, fluxes and oxide films. The requirements for refractories for the vanes of degassing units are the same as for melting and holding furnaces.

4.6.2 ALUMINA–SILICA REFRACTORIES

The aluminosilicates are popular refractories for melting and holding furnaces. High-alumina content (57 to 85% Al_2O_3) refractories, typically consisting of a mixture of corundum (Al_2O_3) and mullite ($Al_6Si_2O_{13}$), are used in areas that come into contact with molten metal and dross. The refractories also contain ingredients like CaO and free silica. The use of pure alumina does not result in improvements commensurate with additional cost.

Silica in any form (crystalline, amorphous or combined) present in the refractory material readily reacts with aluminium. Therefore, the use of aluminosilicate compounds implies the use of remedial measures also to offset this vulnerability. The effective remedial measures are the use of anti-wetting agents and restricting pore sizes to <1 μm to hinder aluminium penetration and decrease the total contact area available for reaction (Bonadia et al. 2006).

When aluminium reacts with the silica and silica-bearing constituents, corundum and elemental silicon are formed.

$$4Al + 3SiO_2 = 2Al_2O_3 + 3Si \tag{4.3}$$

$$8Al + 3Al_6Si_2O_{13} = 13Al_2O_3 + 6Si \tag{4.4}$$

Brondyke (1953) noted that the reaction is accompanied by an increase in volume. Cracks form as a result of this volume expansion. However, Nandy and Jogai (2012) calculated a reduction in volume. Yurkov (2017) also noted that both reactions result in a negative volume effect, which facilitates metal (Al) penetration. In any case, a volume change accompanying a reaction causes stresses within the refractory and degrades its integrity.

According to Brondyke (1953), the aluminium penetration of these refractories proceeds at a decreasing rate because the penetrated material becomes an ever increasing barrier to the diffusion of aluminium through it to the unreacted refractory. As the zone of penetration increases in thickness, the diffusion and hence, the rate of reaction both decrease. Even though penetration of the alumina–silica refractories is the reaction of aluminium with silica and the silica-bearing constituents, Brondyke (1953) discerned no apparent relationship between the percentage of silica (free and combined) in the refractories and the extent of aluminium penetration. It was also mentioned by Allaire (1992) and Afshar and Allaire (1996) that despite the corrosion phenomenon being essentially characterized by silica reduction, no evident correlation exists between metal penetration and the silica content of refractories.

The metallic silicon formed as reaction product ends up in the aluminium bath (Brondyke 1953). But, the silicon pick-up by the aluminium tracks the silica content of the refractory, with a definite decrease of silicon pick-up with increase in alumina content (>50% alumina). Temperature has a significant effect. Penetration of the various refractories was negligible at 705°C but increased rapidly at 760°C and above. The porosity of the refractories apparently had little effect on the degree of penetration, and the penetration proceeded through the refractory grains by diffusion rather than by capillarity action through pore channels. Commercial mullite refractories were penetrated to the same degree and at about the same rate as other refractories having a similar chemical composition without all the silica and alumina being converted to mullite.

Among various investigations on the interaction of molten aluminium with aluminosilicate refractories, the results obtained in apparently similar process conditions generally agree but often do not match closely with one another. This could be due to the inevitable variations in the actual chemical composition of the melt, composition and structure of the refractory.

4.6.3   BELLY BAND ZONE

In the aluminium melting and holding furnaces, the zone most vulnerable to corrosion is the melt line, known as the bellyband (Schlesinger2006). It is shown schematically in Figure 4.7 (McCollum 1989; Afshar and Allaire 1996; Nandi and Jogai 2012). Below the melt line, the reaction between molten aluminium and the silica in the refractory to form corundum and elemental silicon stops because a surface layer of corundum forms. The corundum layer prevents continued contact of the metal with the refractory beneath the layer and stops the reaction. When zinc or magnesium levels are high in aluminium melt, the corundum layer is less adherent, and consequently, refractory corrosion could continue.

The conditions are different at the bellyband. The molten metal penetrates into the pores of the refractory and reacts directly with the oxygen present in the adjacent furnace environment to form corundum. As the temperatures are higher at the belly band, the reaction occurs faster, causing the metal to rise upward through the pores, forming corundum at locations above the melt line, as shown in the figure. The corundum layer is a composite with fine grains of corundum surrounded by an interconnected metallic network. The metallic aluminium network provides the channels to supply fresh aluminium to the reaction interface during the corrosion process. The corundum deposit is strongly adherent to the refractory. Besides, the deposit can grow so large as to even reduce the furnace capacity.

Incidentally, constant or prolonged contact between the melt and the aluminosilicate refractory does not appear to be necessary for such product build-up. Brondyke (1953) noted that build-up also appears on the upper side walls and roof (wherever the aluminium melt would contact the refractory) due to splashing of the melt during charging or gaseous fluxing of the metal.

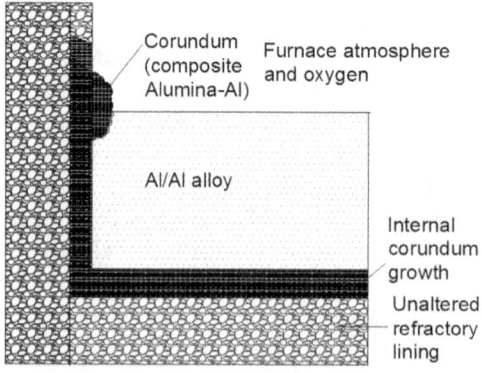

FIGURE 4.7   Bellyband (From Nandi and Jogai 2012, Engel 2015, Schlesinger 2006).

At the metal line, corrosion of the refractory is fast and significant. Afshar and Allaire (1996) noted that the morphology of oxide layers depends upon the oxygen in the environment. Under air, the oxide layer was porous, voluminous and non-protective, while under a nitrogen atmosphere, the oxide layer is thin and more compact. When aluminium contains magnesium, the oxide formed is doped more or less by MgO and is no longer protective. The use of higher-quality refractory and eliminating free silica in the lining were measures recommended by Afshar and Allaire (2001) for reducing corundum formation. A high-density Al_2O_3 ceramic rod displayed very good resistance to corrosion as well as to metal penetration in laboratory tests.

4.7 MANAGEMENT OF ALUMINIUM PENETRATION

The options available to counter the ability of the aluminium to wet and subsequently penetrate and react with the refractory are: (i) the use of anti-wetting additions to refractories, (ii) to use a phosphate bond, because it imparts good non-wetting properties to the refractory, and (iii) the use of sol–gel binders. The additional options (Engel 2015) are increasing the ratio of alumina/silica in refractories, decreasing the content of alkali oxides, magnesium and zinc, and lowering the pore size and the porosity of refractories.

The dimensions of the pore through which the melt cannot penetrate due to the capillary effects vary depending on the temperature, structure and composition of refractories, as well as on the type of melt. The critical pore dimension (below which the melt cannot penetrate due to the capillary effect) for Al melt is <1–2 μm (Yurkov 2017).

4.7.1 ANTI-WETTING ADDITIONS

The use of anti-wetting additions to refractories in contact with molten aluminium is a popular approach, especially for castables containing calcium aluminate. Many different materials have been used for this purpose, and the details remain proprietary. Hence, the exact mechanism(s) of their action remains uncertain (Engel 2015).

The anti-wetting additives are added to the refractory mix during manufacture (Afshar and Allaire 2001). Barium sulphate is the most common additive. Aluminium fluoride and calcium fluoride (AlF_3, CaF_2) are also used (Adabifiroozjaei et al. 2015), as well as certain boron- and phosphorus-containing materials. The decrease of wetting due to anti-wetting additive is not an absolute protection. These additives decompose at a certain temperature (depending on the additive) and once this is exceeded, the refractory loses their protection. Yet, these anti-wetting additives do add years of service life to the refractories.

Ba-based anti-wetting additives include $BaSO_4$, BaO and $BaCO_3$, Ca-based additives are CaO and CaF_2, B-based additives are BN, B_4C, B_2O_3 and $9Al_2O_3 \cdot 2B_2O_3$), and P-based additives are $AlPO_4$ and glass frits. In addition, compounds such as Cr_2O_3, $AlTiO_5$, cryolite, CeO_2, AlF_3 and $ZrSiO_4$ have also been tried.

The anti-wetting additives act by increasing the interfacial energy between the refractory and the aluminium melt through the formation of a new phase or phases

(Adabifiroozjaei et al. 2015) by reactions between the additive and the refractory. Increasing the interfacial energy lowers the tendency of melt to wet the surface of the refractories. Even in non-wetting systems, aluminium and its alloying components tend to diffuse through the interface to equilibrate. Here, anti-wetting materials may result in the formation of dense and rigid interfacial layers that retard the diffusion.

Ba-containing compounds are effective additives. The effectiveness is attributed to the formation of celsian, a monoclinic $BaAl_2Si_2O_8$, in the matrix of the refractory. Celsian forms by reaction between the additive and the refractory components (Al_2O_3 and SiO_2) at temperatures exceeding 1000°C (Afshar and Allaire 1996, 2001). According to Aguilar-Santillan (2008), $BaSO_4$ additions also resulted in a substantial decrease in the number of pores in the refractory.

Afshar and Allaire (1996) emphasize that corrosion of the aggregates was insensitive to the non-wetting additive, and the beneficial effect of the non-wetting agent was present only in the matrix. In some cases, the corrosion of aggregates promotes corrosion in the surrounding matrix, even in the presence of a non-wetting agent (Nandi and Jogai 2012). At temperatures below 1000°C, barium compounds react with SiO_2 and Al_2O_3 to form α-$BaAl_2Si_2O_8$ (orthorhombic phase), which was credited with the protection observed (Schmutzler and Sandhage 1995). Above 1050°C, the orthorhombic phase transforms to a monoclinic phase, and the protection is lost. Incidentally, in their review, Adabifiroozjaei et al. (2015) attributed the non-wetting behaviour to the monoclinic structure of $BaAl_2Si_2O_8$. There are also conflicting statements as to the effectiveness of celsian vis-à-vis barium sulphate. The current information on the anti-wetting action of celsian vis-à-vis barium sulphate remains contradictory. Adabifiroozjaei et al. (2015) consider monoclinic celsian to be conferring superior anti-wetting properties on the refractory, while many other authors consider barium sulphate to be superior.

4.7.2 PHOSPHATE BONDING

A phosphate binder imparts good non-wetting properties to the refractory and does not decompose until temperatures >1500°C (Engel 2015). In brick, it can be incorporated into the mix itself prior to firing. In castables, it can be the bonding agent or supplement it, and in mouldables (plastics), it is the binder. Wollastonite (calcium silicate, $CaSiO_3$) is not wetted by Al and has also been considered as an anti-wetting additive (Vikulin et al 2006) useful below 1125°C.

Generally, the joints in the refractory lining using bricks are particularly prone to melt penetration and attack. Monolithic refractory linings have no joints but are mechanically weaker. Both bricks and monolithic materials are used in furnaces (Nandi and Jogai 2012). Phosphate-bonded materials are known for good resistance to aluminium metal penetration and corundum growth. The addition of non-wetting additives that tend to decompose at temperatures above 1000°C is useful in low-temperature processed materials. Lower firing temperatures create smaller pore sizes and improved non-wetting properties, which translate to higher penetration resistance against alkali.

The use of sol–gel binders, in particular colloidal silica, results in small pores that can hinder aluminium penetration (Engel 2015). In such a situation, a non-wetting component may not be required or may be required in smaller quantities. Certain special raw materials or binders that have low wettability with respect to aluminium contain only lime and alumina, e.g. calcium hexaluminate or bonite. Bonite is a new synthetic, dense refractory aggregate based on the mineral phase calcium hexaluminate ($CaO \cdot 6Al_2O_3$ or CA6) (Buhr et al. 2016; Schnabel et al. 2011). Bonadia et al. (2006) noted that materials based on sintered CA6 or bonite are inherently less wetted by aluminium, do not require anti-wetting additives and keep their properties even at temperatures >1200°C.

4.8 REFRACTORY INCLUSIONS

Silicon oxide, iron oxide, titanium oxide and sodium oxide, which serve as the base and constituents of refractories, are deoxidized (reduced) by Al. In the Al melt, there are always impurities (sodium, potassium and gases) and alloying components (e.g. magnesium, strontium, zirconium). The corrosion of aluminosilicate refractories (in melting and holding furnaces) by molten aluminium generally leads to the formation of an alumina deposit on the refractory. Such corrosion also promotes the formation of inclusions in the molten metal by erosion. Inclusions can also originate from direct oxidation of impurities in aluminium at the metal line (Allaire 2000).

The products of oxidation are a function of the alloy composition. At operating conditions of aluminium melting furnaces, the presence of 0.3 to 18 wt.% magnesium in aluminium favours the formation of spinel ($MgAl_2O_4$). At less than about 0.3 wt.% Mg, alumina (Al_2O_3) is the most stable oxide in these conditions, while at more than 18 wt.%, it is magnesia (MgO). These oxides may result from the action of oxygen gas on the molten metal, which leads to the formation of a skim at the metal line. This direct oxidation may also occur at the refractory/metal interface due to the oxygen gas that migrates through the pores in the refractory. The oxidation of aluminium by less stable oxides such as silica occurs in the more reactive regions of the refractory, where vitreous phases surround more stable crystalline phases. The latter are released to form inclusions in the metal.

Alkalis contained in the refractory (usually <2 wt.% Na_2O) are highly susceptible to reduction by aluminium, and sodium is released. AlF_3 in the refractory, as a non-wetting agent, also contributes to the release of alkalis into the molten metal. Sodium in the molten metal oxidizes, and Na_2O converts the protective alumina layer into sodium aluminate ($NaAlO_2$), whose density (2.69 g/cm^3) is lower than that of alumina (3.96 g/cm^3). Because of this difference in density, the protective alumina layer cracks, and melt penetration is facilitated.

4.9 REFRACTORIES RESISTANT TO MOLTEN ALUMINIUM

Aluminium oxide, magnesium oxide, barium oxide and calcium oxide are refractories stable to molten Al. Magnesia bricks were used for the lining of furnaces in the 1960s, but the use did not continue due to their low thermal shock resistance. Yttrium oxide and zirconium oxide also may be resistant to molten Al. Aluminium nitride is completely resistant to aluminium even at 2000°C.

Sialon was tested in molten aluminium. The corrosion test was performed in an electric furnace at 800°C for 150 h in static air. In the experimental arrangement, the sialon specimen was in contact with (i) air and molten aluminium simultaneously, and (ii) only molten aluminium (no contact with air). Scanning electron microscopy (SEM) studies showed no traces of Al in the pores of the crucible. There was no infiltration. The material showed excellent resistance in contact with molten aluminium evenafter extended exposure (150 h) at 800°C. It has been suggested that formation of AlN scale at the contact point of Al(l) and sialon is responsible for the very slow corrosion attack of sialon and silicon nitride ceramics (Brook 1991), a tenable explanation in an oxygen-free atmosphere, as it occurred at the bottom of the crucible, floating inside the Al melt.

4.10 COPPER

Mark Schlesinger (1996), in his review on the refractories for copper production, indicated that the furnaces used for producing molten copper from concentrates and scrap – flash smelters, converters, and anode and firerefining furnaces – face exemplary challenges to the refractory life in the form of highly aggressive slags, mechanical stresses, batch operation and higher operating temperatures. When everything appeared to be settling down, the use of 'mag–chrome' refractories began to be scrutinized and questioned as being a source of a possible environmental villain, hexavalent chromium. In a way, the search for a 'more' suitable refractory begins all over again.

The copper production flowsheet consists of (i) matte smelting, in which sulphide-based copper concentrates are melted and partially oxidized to forma silicate slag and a molten matte. The matte is animpure mixture of sulphides consisting primarily of copper and iron sulphides; (ii) converting,where oxygen-enriched air interacts with the molten matte to produce molten blister copper and a slag containing most of the iron; (iii) fire refining, where the blister copper is partially refined in 'anode' furnaces and cast into anodes, and (iv) electolytic refining, where relatively impure copper in the anode is electro refined and recovered as pure copper cathode. There are also refining furnaces for melting and removing impurities from scrap copper.

4.10.1 Smelter

For matte smelting, as well as reverberatory smelters, electric, submerged tuyere and flash furnaces are in use. The sulphide mineral concentrates are melted down, and part of the iron and sulphur in them is oxidized. The products of smelting, coming from concentrate and from the burning of carbonaceous fuels, are: (i) a gas stream containing SO_2, N_2, CO_2, H_2O and O_2; (ii) fayalite slag, Fe_2SiO_4 (m.p.1200°C), formed by the reaction of silica in the concentrate and added as a flux, and FeO, formed on oxidation of iron sulphides. The slag also carries the gangue oxides, Al_2O_3 and MgO, and also some (~2%) copper oxide generated during the oxidation reaction; (iii) the matte – a molten solution of Cu_2S, FeS and minor amounts of sulphides of lead, nickel, tin, zinc, cadmium and other elements; and (iv) the dust-laden off gas. Substantial amounts of copper, arsenic, iron, bismuth, zinc and antimony, besides a few more elements, are carried off in this stream.

In the reverberatory and flash smelting furnaces, as well as in the Peirce–Smith and top blown rotary converters, there is chemical, thermal and mechanical attack on the refractory. FeO is an effective flux, and that makes the FeO-laden basic slag the main factor. In flash smelters, slag lines the reaction shaft, and near the top, because of prevailing high oxygen partial pressure, FeO oxidizes to magnetite, which is good because the solid magnetite coating protects the lining underneath. Near the bottom, such conditions do not exist, and no protective magnetite forms. The lining in this portion is more vulnerable to slag corrosion. In the flash furnace, the falling droplets of matte and slag are at a higher temperature and thus can dissolve and wash out the refractory. In reverberatory furnaces also, the use of oxygen-enriched air pushes up the temperatures in the roof and side walls and makes the environment difficult for the refractories. The mixture of water vapour and sulphur dioxide in smelter off gas is deleterious to the basic refractories, because they can be hydrated by water vapour, and SO_2 can convert them to fusible sulphates.

4.10.2 CONVERTER

In the converter, matte is converted into blister copper and slag, and the conditions for the refractories are harsher here. Peirce–Smith (PSC) units have been almost exclusively used to convert all the primary copper produced. Enriched air blown through the tuyeres in the side of the vessel, in the first stage of the blowing process, oxidizes the iron sulphide in the matte, and the iron oxide formed reacts with the added silica to form fayalite-based slag. The nearly pure copper sulphide remaining is oxidized in the second stage to blister copper and slag. In the second-stage blow, a good amount of Cu_2O is also formed and collects in the slag, and this slag is reverted back to the smelter. In the end, there is still about 0.2% S remaining in the blister copper.

Cuprite or Cu_2O is a highly effective flux, and its presence makes the molten slag very corrosive. Batch operation always implies deep temperature cycling and mechanical impact due to periodic charging of the converter as well as punching of tuyeres to unclog them. The refractories thus have to face mechanical damage and thermal stress components. Refractory erosion by molten phases becomes more intimidating due to bath agitation by submerged gas blowing (PSC) as well as by complete rotation of the vessel (Top Blown Rotary Converter or TBRC). The infiltration into the furnace bottom and virtual freezing of molten copper is another hostile occurrence special to converters.

The converter yields blister copper, which contains various metallic and non-metallic impurities, and many of them are removed in a subsequent process known as fire refining. In fire refining, less noble elements like arsenic, antimony and lead are preferentially oxidized and removed from molten copper using a basic flux. Fire refining is carried out in hearth and rotary furnaces. A high concentration of cuprite also accumulates in the fire refining slag because of the highly oxidizing conditions used. The melting of scrap copper is also carried out in such furnaces. The second part of the fire refining process and the final pyrometallurgical step in a typical copper production flowsheet is the deoxidation and purification of molten metal using hydrocarbon gases or occasionally, by the ancient practice of using wooden poles. The slags in the refining furnaces are more basic than those in smelters and converters and have a much higher Cu_2O loading.

Fire refining furnaces are operated on a semi-continuous basis and thus, have less thermal shock issues. Secondary copper smelters, like converters, are batch operated with all the attendant load on refractories. These are usually rotary furnaces, where the refractory life is shortened by erosion and mechanical stress as well as chemical factors. Converters that (i) operate continuously and (ii) do away with furnace motion or rotation, which would ease thermal shock and mechanical stress factors on refractories, are still some distance away from routine deployment.

4.10.3 THE MAG–CHROME REFRACTORY

Flash smelters dominate the copper smelting facilities. Reaction shaft and slag line are two areas prone to refractory wear and so, are lined with fused grain chrome–magnesia bricks. Lower MgO compositions (less basic) are chosen because the SiO_2 and/or water vapour in the smelter off gas would attack the MgO grains in higher-basicity bricks. In areas less prone to slag attack, such as off gas shaft and side walls, direct-bonded chrome–magnesia refractories are fine.

Silica and alumina refractories were still in use until the 1990s for roofs and side walls of reverberatory furnaces. These areas do not come into contact with the slag and matte, and these refractories can cope with the presence of SO_2 and water vapour in the environment. Later, when oxy-fuel burners were positioned in the roof, the roof temperatures became higher. Because of this, and to deal with the corrosive impact of dust and fume emanating from the smelting process, mag–chrome bricks, either burned or directly bonded, began to be used. Beginning in the 1950s, converters have been exclusively lined with mag–chrome. The best known (available) material is used to line the tuyere area, where the highest refractory wear occurs. Originally, burned brick was used. It is now direct-bonded refractory with fused grain (Rigby 2005).

The bricks face penetration by matte and fayalite slag initially, and later, by copper oxide. The use of enriched oxygen leads to higher temperature at the tuyeres and exacerbates thermal shock, consumes the protective magnetite layer and increases the solubility of the refractory components in both first (iron blow) and second (copper blow) slags (Schlesinger 1996). Any extra blowing to eliminate hard-to-oxidize lead and antimony from the blister copper generates more copper oxide,which penetrates the refractories and causes failure by slabbing.

When high-magnesia materials are used in the tuyere area, they react with the fayalite slag generated during iron blow, forming a grain boundary solid solution of fayalite and forsterite, which limits the subsequent penetration of matte and copper oxide. A highly siliceous slag ends up dissolving this solid solution,exposing the spinel grain, and thus promotes both penetration and refractory loss by erosion.

High-chrome compositions (it now becomes chrome–magnesia) better resist lower temperature corrosion by molten silicates and are used throughout the converter. The reaction between cuprite in the slag and chromic oxide in the refractory grain generates solid copper chromite ($CuCrO_2$), which coats the refractory surface. Further damage is then limited.

Chrome–magnesia bricks are also favoured in TBRCs due to their high strength as well as other properties. Lower brick porosity reduces slag penetration. The reduced thermal shock resistance associated with lower porosity is not critical away from the tuyere zone.

In anode or fire refining and secondary smelting furnaces with high slag basicities, the absence of siliceous slags points to the use of highly basic (magnesite) refractories. Even here, mag–chrome refractories are widely used. The most corrosive environment in anode furnaces is the slag line. The combination of high cuprite slag and air or oxygen here gives rise to the same melt penetration situation that impacts the refractory life in the converters. Fused grain and fused cast mag–chrome bricks need to be used in this region. At other places, direct-bonded bricks are good enough.

Oxidized copper penetrates the refractory easily. Thus, tar-impregnated magnesite bricks are used in the floor. The reducing environment created by the presence of tar reduces the dissolved oxygen content in the molten copper, which in turn, reduces its ability to penetrate the refractory itself. Once infiltrated, the copper oxidizes to form solid Cu_2O and CuO; the brick grows and eventually fails by bursting (spalling). In an oxidizing environment, the likely formation of a low-melting-point eutectic in the Cu_2O–MgO system precludes the use of magnesite refractories for copper containment.

In a typical new-generation copper converter, which is stationary, the turbulence created by the vertically injected oxygen is used to achieve the mixing action produced by the horizontal tuyere injection of PS vessels or the rocking rotary action of TBRC. Mechanical stress on the refractories is thus less. Because the off gas generated in the new converters is richer in SO_2, the use of basic refractories becomes difficult. Impurities removal from high-grade mattes in newer converters will be less effective, and the presence of impurities in blister copper can degrade refractory performance. Besides, the greater reaction intensity in the newer converters coupled with higher oxygen enrichment will result in higher temperature in the reaction zone.

Beginning in 1990, all solid wastes containing chromium oxide, including mag–chrome bricks, came to be treated as hazardous material. The concern was over Cr^{3+} in these materials,which might be oxidized to Cr^{6+}, and Cr^{6+} is water soluble. Cr^{6+} formation in refractories results from reactions in oxidizing environments with alkaline components (Na_2O, CaO, etc.) in slag or in the brick itself to form chromates. In current practice, the chemical environment used in the smelters and converters does not favour the formation of hexavalent chromium in copper production (Schlesinger 1996). The use of basic fluxes is generally not very prevalent. However, to improve removal of arsenic and antimony, some refining practices do use sodium carbonate or hydroxide. The use of calcium ferrite slags in the converting of high-grade mattes may also result in the generation of Cr^{6+} in refractories.

The way out, then, is development of a chrome-free refractory suitable for replacement of mag–chrome bricks in copper production furnaces. The clue is the MgO–Al_2O_3 phase diagram, whichis similar to the MgO–Cr_2O_3 phase diagram.

Refractories based on mixtures of MgO and Al_2O_3 could be suitable replacements for mag–chrome (Schlesinger 1996).

Magnesia-alumina refractories may be as effective as mag–chrome, if not better, in copper production furnaces. The solubility of $MgAl_2O_4$ in molten copper oxide at 1400°C is slightly lower than that of $MgCr_2O_4$. Intergranular penetration of oxidized copper into magnesite refractories can be equally well inhibited by the addition of Al_2O_3 or Cr_2O_3 to the brick composition. In calcium ferrite slags, MgO with 20% added alumina corroded at about the same rate as magnesite with 20% added chrome. Mag–alumina instead of mag–chrome in almost every copper production application is a possibility. In smelters, periclase–spinel brick might have better resistance to the slag penetration. Alumina-rich compositions have better resistance to sulphation and might be useful in roof and stack locations.

Steel ladle slags and flash smelter slags are similar. Alumina-rich compositions working well for slag linings might be useful for flash smelters as well, but the resistance of mag–alumina refractories to attack by high-cuprite slags needs to be checked and validated. With increasing use of calcium ferrite slags in converting operations, periclase–spinel or spinel–periclase might play a wider role.

Mitsubishi introduced the use of a calcium ferrite slag for converting. In comparison to fayalite, calcium ferrite slag has a lower viscosity and also a lower solubility of Cu_2O. Its higher basicity is conducive to the removal of impurities that form acidic oxides. However, due to the low solubility of silica in this slag, the use of this slag in smelting processes is restricted. Aided by the low viscosity of calcium ferrite, magnesia–chrome bricks could undergo significant refractory penetration and wear (Fahey et al. 2004). The Ca-ferritic slag corrodes not only the basic magnesia component but also the chromite and forms low-melting Mg-Feoxide (magnesia–wuestite).

4.10.4 CALCIA–MAGNESIA

In nature, calcia and magnesia occur as the mixed carbonate dolomite. On calcination, dolomite forms an oxide mixture of periclase and lime or doloma. It contains about 60% CaO. The manufacture of refractories from this had to use additives like iron oxide, silica or alumina to produce a lower-melting bonding phase. The temperatures required for direct bonding are impractically high for any manufacturing facility. The addition of binders is also needed to improve the density of the fired product, since doloma hydrates easily. Tar or pitch can also be used as a binder instead of acidic oxides. The tar has multiple roles: it not only minimizes hydration during storage but also serves as a reducing agent in contact with molten metal. For instance, the removal of dissolved oxygen from copper blunts its ability to penetrate grain boundaries.

Oxidation of the matte generates SO_2, which can permeate into the brick. The SO_2 subsequently oxidizes to SO_3, which in turn, reacts with the basic oxides of the chrome–magnesia brick to form the alkaline earth sulphates, $MgSO_4$ and $CaSO_4$, at temperatures below 1050°C. These reactions are associated with volume increase, which also causes filling up of the brick pores. The densification of the refractory structure and the weakening of the brick bond eventually cause cracks in the bricks, facilitating melt infiltration.

If metallic copper permeates into the refractory, there are two consequences. One is an increase in thermal conductivity of the brick and as a consequence, substantially higher temperatures at greater depths of brick. This will affect both corrosive and thermal reactions. Incidentally, pure copper scarcely infiltrates the brick; oxygen contamination facilitates penetration through its effect on the contact angle between Cu and the refractory oxide.

Another possibility is mixing pure magnesite with doloma during refractory production, generating a mag–doloma brick analysed at 60% total MgO. Mag–doloma bricks hydrate less easily and resist slag corrosion better than pure magnesia. Doloma and mag–doloma might be viable options for refining furnaces with their highly basic slags and highly oxidized copper. The use of dolomitic refractories in BOF linings subjected to similar temperatures suggests that these materials would be competitive.

Malfliet et al. (2014) have summarized some consequences of melt infiltration into refractories used for copper production. Melt infiltration occurs due to the penetration of molten slag, copper sulphides and oxides, and the reaction products that are formed. The acidic character and low viscosity of the slags, high temperatures and the presence of additives such as soda aggravate the slag corrosion and make it one of the main wear mechanisms of refractory in the copper industry. The slag dissolves specific phases in the bricks' microstructure.

Slags in the primary smelting and converting processes are of the ferrous silicate type. On contact with fayalite slag, refractory corrosion occurs through the dissolution of MgO, more so when the silica content of the slag is higher, the reason being that periclase (MgO) is more basic than chromite and therefore, more susceptible to acidic corrosion. The chromite grains do not react with the slag but are washed away once they are unbound. Refractories that are higher in chrome ore and are direct-bonded with a spinel bond rich in Cr_2O_3 are more resistant to chemical corrosion.

A high SiO_2 supply caused by changes in the processing and/or the uncontrolled addition of silica sand results in the copious formation of forsterite (Mg_2SiO_4) following contact with periclase, which due to the associated volume expansion, causes forsterite bursting and deterioration of the brick structure.

4.10.5 OLIVINE SLAGS

A slag with properties midway between those of fayalite and calcium ferrite slags is the ferrous calcium silicate or olivine slag used in the Ausmelt C3 converting process. Yazawa (2001) proposed this slag to reduce the copper loss, increase the absorption of both basic and acid oxides, and limit refractory problems.

4.11 LESS COMMON METALS

The unique technology benefits of a large group of metals known collectively as rare metals or less common metals could be accessed only when methods were developed for their production in sufficient purity and for processing in good quantities. Reflecting their unique nature or potential application areas, groups among them have been referred to as reactive and refractory metals, nuclear metals and rare earth metals. These metals were little more than laboratory curiosities but emerged into prominence in the post-World War II period, and their development was driven by the nuclear, aerospace, electronics and biomedical industries. Interestingly, even today, these metals remain less available and less familiar to metal users at large. They are notoriously reactive and need elevated temperatures for their production and processing, and the task of handling almost every member of the less common metals group has been filled with challenges. The process for the production of many of these metals is metallothermic reduction, also known as metal replacement reaction. This is dictated by sound science and technology reasons as well as cost considerations.

4.12   METALLOTHERMY

In contrast to the ferrous as well as the common non-ferrous metals, the less common metals are prepared not from beneficiated ores or concentrates but from inter-process intermediates, which are invariably a pure compound usually obtained after elaborate separation procedures. The objectives of the metallothermic reduction have been summarized by Herget (1985) as follows: (i) the reaction should occur quickly and give product metal in high yields, (ii) the reaction products must be obtained in compact forms – metal as an ingot and the slag as a well-separated layer, (iii) the reaction should be self-sustaining and once initiated, should proceed without the need for additional external heating, (iv) the product metal must be of high purity, (v) the reaction proceeds in an open atmosphere, (vi) the reaction should proceed safely without risk, and (vii) the reaction should be amenable to being carried out in commercially available reactors using readily available refractories as containers. Metallothermic reductions are generally exothermic reactions, and by a suitable choice of reactants, fluxes and fluidizers, and process conditions, it has been possible to realize many of the listed characteristics to a great extent and obtain the metal from its compound. To repeat, refractories have played a key role in the implementation of metallothermic reductions.

For a host of technical reasons, the compound chosen is either an oxide or a fluoride or a chloride. Other things (such as reducibility, purity of as-reduced metal and possibility of post-reduction purification) being equal, an oxide is a convenient choice and invariably preferred. Between the fluoride and chloride, the chloride is the more versatile intermediate but is difficult to handle because of its volatility, hygroscopicity and corrosive nature. Fluorides undergo a neat reduction but do lack certain advantages special to chlorides. All these compounds, with very occasional exceptions, are reducible under technologically useful conditions by metals such as aluminium, calcium, magnesium, sodium, potassium and lithium. Reduction using a metal invariably needs a high temperature, attained either by external heating or by the exothermicity of the reaction or by a combination of both. Thermal management of the process is an integral part of using metal as the reducing agent. Besides,

with very few exceptions, all these reductions have to be carried out under a protected atmosphere (vacuum or inert gas) with inert containment arrangements to
protect the compound intermediate, the reducing agent and the as-reduced metal
from contamination.

Metallothermic reduction has been used to obtain metals in a variety of physical
forms, from well-consolidated ingots to a fine powder. If the heat generated during
the reaction is sufficient to raise the temperature of both slag and metal above their
melting points, and if they remain molten for a sufficient length of time, so that
the molten metal (denser) settles by gravity with the immiscible slag layer (lighter)
remaining on top, a metal ingot topped by a solidified slag results on cooling. Slag
and metal are then separated mechanically. When the heat of reaction is insufficient
to result in an all-liquid reaction mixture, other types of products can form. If the
reduction occurs, but neither the metal nor the slag melts, the metal is formed as a
powder dispersed in the slag matrix. The slag in this case is leached away to separate the metal powder. Metal in the form of powder can result even if the slag melts
when the melting point of the metal is very high and the metal is not sintered to
any extent at the temperatures involved. When the reduction to metal occurs with
only the slag melting, but not the metal, and the melting point of the metal is not
very high, the metal particles may coalesce and undergo partial consolidation into a
sponge. Here, the metal and sponge are separated either by aqueous leaching or by
vacuum distillation.

By a suitable choice of reductants, reactors and reaction conditions, the metal
can usually be obtained in the chosen form. The form of the product obtained in
the reduction determines how difficult it is to select a suitable container or crucible.
For example, the notoriously reactive metals titanium, zirconium and vanadium are
obtained in the form of sponge at the primary reduction stage when their chlorides
are reduced with magnesium at temperatures lower than 1000°C. The crucibles
usable for reduction are mild steel or Inconel. If the reduction process were to involve
handling these metals in a molten form, choosing a crucible would become a com-
plicated and almost unsolvable problem. The seriousness of a reactive melt increases
first of all with its reactivity and also when the melt forms at a higher temperature,
in larger quantities, and persists for longer durations. The formation of the metallic
melt and its proper containment are invariably necessary when the metal in the as-
reduced condition is obtained in the form of an ingot.

The properties of reactants and products, temperatures reached during reduction,
scale of operation and impurities tolerated in the as-reduced metal – all these determine the materials of construction of the reactor and the choice of melt facing refractories, be it a metal or a compound. In the as-reduced condition, less common metals
are obtained in the form of an ingot in two types of reactors. One is the bomb reactor,
which is usually a refractory-lined steel vessel that is closed during the course of a
reduction after loading the charge. Calciothermic and aluminothermic reduction of
oxides and calciothermic and magnesiothermic reduction of fluorides are carried out
in bomb reactors.

In many metallothermic reductions, an adiabatic environment is usually arranged
for a certain duration when the objective is to obtain the product in the form of an

ingot. This is achieved by selecting a reaction that occurs rapidly and providing good thermal insulation of the system. The total heat input within the reaction volume through exothermicity of the main reaction, exothermicity of the booster reaction and external heating (usually as a means of initiating or triggering the reaction) is carefully conserved and utilized in three ways. The first is the choice of a reactor that is thermally insulated, such that the heat once accumulated during the reaction is not lost from the reactor volume before slag–metal separation is complete. A booster reaction is a reaction of high exothermicity (high kJ per gram of reaction mixture) that occurs alongside the principal reaction, independently but in the same reaction mixture. It shares the reducing agent, and the product should flux with the product of the main reaction to form a fusible and/or fluid slag. Heat is best conserved in closed bomb reactors, which are essentially steel vessels lined on the inside with calcia, magnesia or alumina refractory The refractory lining is principally to contain the melt while it forms and separates into two layers – metallic melt as the bottom layer and slag as the top layer. The lining resists and is inert to the attack by the melt constituents, and that is why it is there in the first place. It also stops any rapid loss of heat by conduction because it is not a good thermal conductor. Second, metallothermic reductions, particularly the thermit types, occur very fast. Once triggered, they take a minute or two to go to completion even for a large batch size. Sufficient heat for an intended metal–slag separation is generated fast and uniformly throughout the reaction mass. With a fairly well-insulated system, almost all the heat generated persists there for molten slag and molten metal to clearly separate into two layers. Third, the density difference between the metal and the slag is usually large and conducive to a rapid separation. This is further facilitated by ensuring that the slag is sufficiently fluid by itself or made sufficiently fluid by using slag fluidizers. These are the general features of a metallothermic reduction process for the production of less common metals in the form of an ingot.

Calciothermic reduction of fluorides yielding the reduced metal in a consolidated form is also carried out by reducing many of the rare earth trifluorides in an externally heated tantalum crucible under an inert gas cover. In these cases, the heat of reaction is not a major factor in meeting the thermal requirements of the process.

4.13   REDUCTION OF URANIUM TETRAFLUORIDE

Much of the systematic research on metallothermic reduction for ingot metal production was done on uranium, and the results obtained proved to be highly useful leads for various other metals as well (Wilhelm 1955, 1960; Wilhelm and Peterson 1961).

Uranium is a highly reactive metal and forms stable compounds with a number of non-metals. The flowsheet of uranium does not normally involve extensive post-reduction purification, beyond vacuum melting. It is therefore necessary that the metal in the as-reduced condition is pure.

A lining of dolomite was originally used, but a thick lining of calcium or magnesium fluoride can be used as the refractory lining if it will not collapse through melting when in contact with molten slag of the same composition. Calcium fluoride, generated as a product of the calciothermic reduction of uranium tetrafluoride, was routinely used as a lining in uranium metal reduction reactors. The lining prevents the contact between uranium melt and the steel crucible, and using the slag itself as lining eliminates the possibility of ingestion of impurities into metal from other lining compounds. This may seem similar to the popular process of slag coating of BOF vessels. The idea ofusing the slag for furnace lining appears to be much older. Centuries ago, moulded slag blocks were used to build the shaft of Swedish iron-making furnaces at Soderfors.

After placing the charge ($UF_4 + 10\%$ excess Ca) into the lined reactor, a reaction is initiated from cold by electrically heating a small coil of molybdenum or tungsten embedded in the charge or trigger mixture ($Mg + Na_2O_2$) placed in the charge. An arrangement is shown in Figure 4.8. The run, which is over in minutes, yields a clean ingot underneath a solidified layer of slag. It is recovered without necessarily destroying the lining. The CaF_2 slag obtained in the process is crushed, ground, screened and reused for lining the reactors in the next heat. The reactor may be cooled by circulating water through the cooling coils brazed on the outer wall of the reactor vessel.

Calcium reduction may also be carried out in an open bomb and the charge ignited at room temperature. A protective cover of calcium vapour backed by a CaF_2 layer prevents the product metal from contacting the air for oxidation. Uranium ingots thus obtained are usually remelted in a graphite crucible in a vacuum induction furnace and cast into cleaner ingots.

Both in operation and in the quality of the metal produced from it, calcium reductionis entirely satisfactory. However, it has been largely replaced by the lower-cost magnesium reduction process, which gives an equally good product.

In the technique of magnesium reduction of uranium tetrafluoride developed in USAEC (United States Atomic Enegy Commission) cylindrical reactors lined (25

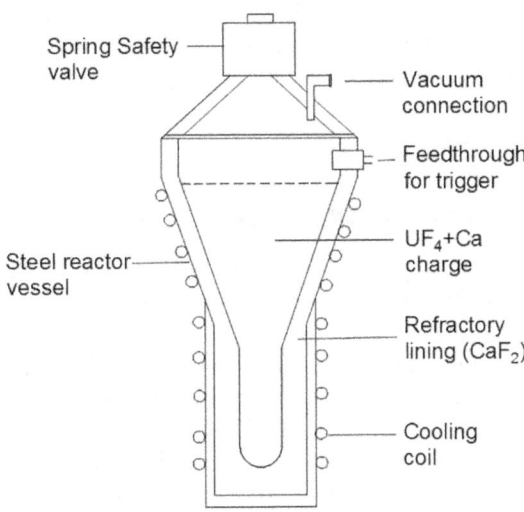

FIGURE 4.8 Reactor for uranium tetrafluoride reduction with calcium.

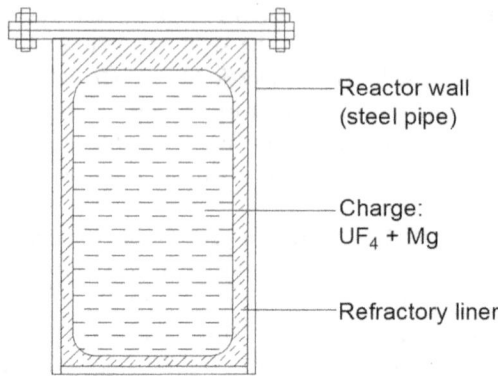

Reactor wall
(steel pipe)

Charge:
UF_4 + Mg

Refractory liner

FIGURE 4.9   Closed bomb reactor for uranium production (From Wilhelm, H.A. 1960).

mm thick) with electrically fused dolomite are used (Jamrack 1963). (Burned lime
and MgF_2 are also suitable lining materials.) The reaction is initiated by heating the
entire assembly in a gas-fired furnace at 365°C for about 3 h. Unlike calcium reduction, the option of carrying out the reduction in an open bomb does not exist for magnesium, because apart from limited exothermicity, the boiling point of magnesium
(1091°C is lower than the melting point of uranium (1132°C). It has to be a closed
reactor. A schematic is shown in Figure 4.9.

Graphite is a suitable lining for a magnesium reduction process. Graphite tends to
react with calcium, and the product ends up with the carbon impurity. Problems due
to high thermal conductivity and specific heat are circumvented by using only a thin
graphite lining, separated from the reactor wall by a gas space or a loose packing
of insulating powder (reminiscent of the composite lining – barrier refractory and
insulation layer – used in steel and aluminium metallurgy). The process is executed
like a typical magnesium reduction. The ingot, slag and graphite come out separated
fairly cleanly. No significant metal–crucible interaction happens, and the graphite
assembly is largely reused.

4.14   REDUCTION OF THORIUM TETRAFLUORIDE

Thorium melts at 1750°C, approximately 620°C above the melting point of uranium,
and is the highest -melting component of the reaction mixture. Calcium fluoride
melts at 1418°C. Sulphur is used as the booster with a large excess of calcium. The
reaction between calcium and sulphur is highly exothermic, and the heat evolved is
available for meeting the thermal requirements of the main reduction process. The
bomb-type reactor is lined with graphite. The reactants are compacted together to
form pellets and stacked inside the graphite liner. The reaction is initiated by means
of an electrically heated molybdenum coil embedded in a trigger mixture of a few
grams of calcium and sulphur. Well-formed thorium billets are obtained in over 90%
yield. This was the UK process. Average impurity contents of billets produced were

as follows: oxygen 0.26%, nitrogen 0.038%, carbon 0.24%, sulphur 1.0%, fluorine 0.33%, calcium 0.21%.

In the American process, the reaction takes place in a closed steel vessel, which is lined with electrically fused dolomite or lime. The charge mixture (ThF_4, $ZnCl_2$, Ca) is designed to yield Th–6%Zn alloy, which melts at 1200°C. Post reduction, the zinc is distilled off from the Th–Zn alloy under vacuum between 1000°C and 1100°C in graphite pots. The development of the process, focused on keeping the reaction temperature as low as possible, concurrently eliminated crucible or lining reactivity with thorium from developing into an issue.

4.15   REDUCTION OF OXIDES

The thermit process, invented by Hans Goldschmidt over a century ago (Goldschmidt and Vautin 1898), has been the framework for all future commercial processes for the production of vanadium and niobium metals and their alloys as well as numerous low-carbon ferroalloys.

The first ingot of vanadium metal was prepared by a process involving calcium reduction of vanadium oxides. No calciothermic reaction for ingot production aims to melt CaO (m.p. 2570°C) unaided. Iodine and sulphur are highly suitable for boosting calcium reduction of vanadium oxides. Not only does the heat thus released add to the heat released by the oxide reduction, but the product, calcium iodide or calcium sulphide, fluxes the main CaO slag, there by reducing the temperature needed for a molten slag.

In 1954, Beard and Crooks carried out the calcium reduction process on a kilogram scale in a reactor similar to that shown schematically in Figure 4.9. A tapered magnesia crucible was placed inside the steel crucible to serve as the liner. Vanadium ingot produced by this method was about 99.7% pure, and the yield of metal was 84%. Oxygen and nitrogen contents were about 0.03% and 0.01%, respectively. In a later report, Chindgren et al. (1963) noted that magnesia crucibles tended to contaminate the vanadium metal obtained with iron and silicon. The vessel lining was changed to one formed in place by tamping magnesia powder, relatively iron and silica free, around a tapered wooden mandrel that served as a form. Such a lining also served as insulation against excessive heat loss and protected the reactor vessel (steel) from attack by molten vanadium and slag. This method of lining the bomb reactor was generally the standard in all subsequent bomb reduction experiments everywhere.

Sodium is not generally used as the reducing agent due to its rather low boiling point (880°C), and potassium is too expensive. Magnesium and calcium are favoured. Magnesium is the cheaper of the two, but reactions with magnesium also form an extremely high-melting oxide, MgO (m.p. 2852°C), as slag. The melting point of calcia is also high at 2572°C. Aluminium melts at a relatively low temperature (660°C), and the boiling point is fairly high (2470°C). The oxide formed during reduction, Al_2O_3, melts at a relatively low temperature for a refractory oxide (m.p. 2072°C). Besides, alumina can be readily fluxed with CaO and CaF_2, and the molten slag can be made very fluid.

Reactors for metallothermic reduction have usually taken simple geometrical shapes and are made from or lined with a small selection of refractory materials: lime, magnesia, electrically fused dolomite, calcium fluoride and graphite. On another level, a water-cooled copper vessel has also been used as a reactor for metallothermic production of ingot metal/alloy. The third category of reactors is simple crucibles, usually of refractory metals, used in a controlled environment. They are mainly, or almost exclusively, used for the preparation of rare earth metals.

Even though early investigations on the preparation of consolidated vanadium relied on calciothermic reduction of oxides, eventually, aluminothermic reduction turned out to be more convenient not only in the reduction stage but also for post-reduction purification. The seminal work on aluminothermic reduction of Group V metal oxides was published from Ames Lab in the same year, 1966, for both vanadium and niobium (Carlson et al. 1966; Wilhelm et al. 1966).

In the experiment described by Wilhelm et al. (1966), the Al_2O_3 slag was discarded; however, it could very well be used in subsequent bomb reductions as crucible liner material. As in the case of aluminothermic reduction of vanadium pentoxide, no booster or slag fluidizer was used in the case of niobium reduction either. The reactor was essentially the same as that used for vanadium – steel shell, alumina lining and closed (Figure 4.10).

The simplest working arrangement in aluminothermic reduction is exemplified by the process known as open aluminothermic reduction for the production of niobium metal. A schematic of the reactor appears in Figure 4.11 (Kamat and Gupta 1971)

The reaction was triggered, with the entire charged reactor at room temperature, by electrically heating a nichrome coil embedded in the centre of the charge. The reduction is very intense, with good evolution of white fumes, and is over in

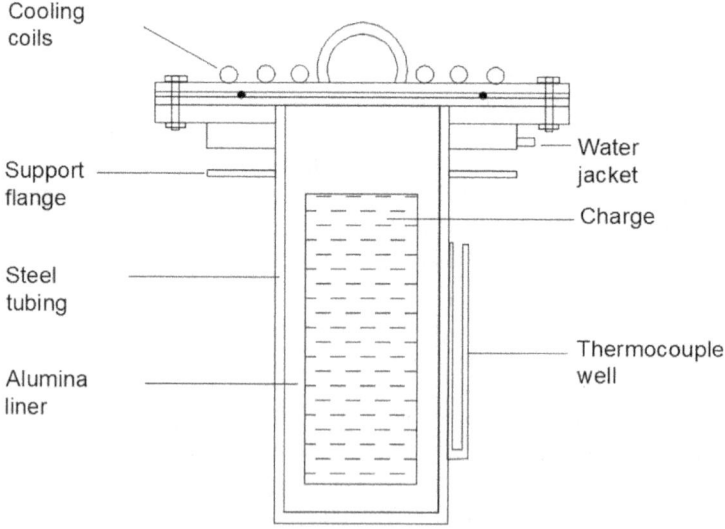

FIGURE 4.10 Bomb reactor for vanadium and niobium (From Carlson, O.N., Schmidt, F.A., and Krupp, W.E. 1966).

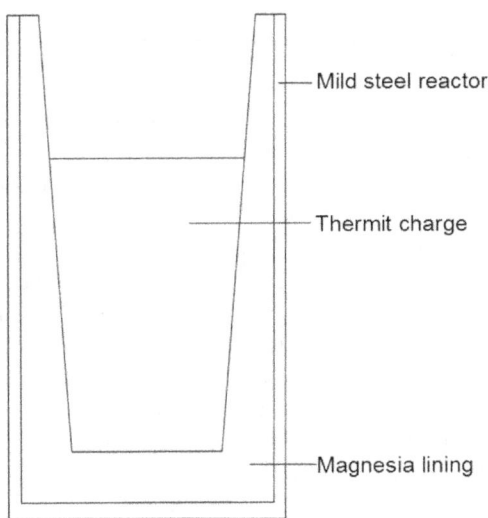

Mild steel reactor

Thermit charge

Magnesia lining

FIGURE 4.11 Open reactor for aluminothermic reduction (From Kamat, G.R. and Gupta, C.K. 1971).

minutes. It was usually possible, immediately (say 30 seconds to a minute) after triggering the reaction, to shovel more charge (Nb_2O_5–Al mixture) into the reactor to continue the reduction and more fully utilize the reactor volume. After overnight cooling, the thermite (Nb–4Al–0.3O in wt.%) topped by an alumina slag is easily recovered from the reactor (invert and lightly tap).

It may be recalled that the great exothermicity of the UF_4 + Ca reaction permitted the production process to be carried out in an open reactor also. The evolving fumes and slag cover protected the molten metal from any atmospheric contamination. The exothermicity could afford the additional loss of heat that happened due to the open reactor and still keep enough heat for slag–metal separation. The situation in open aluminothermy is somewhat similar. The exothermicity of Nb_2O_5–Al is good enough for the reduction reaction in an open reactor. Adequate slag–metal separation was still occurring. The oxygen contamination in the thermit was also not excessive. Metal yield was also good at 98%. The only significant difference is that use of 15% excess aluminium in the charge resulted in only 4% aluminium loading in the thermit. In a closed bomb reduction, the corresponding figures were 5% and 2%. Some of the aluminium undergoes atmospheric oxidation and is lost in the open reduction. This loss is avoided in the closed bomb.

Vanadium can also be produced smoothly by open aluminothermic reduction. Mukherjee and Gupta (1972) ventured to use this process in a sequence that would end up with pure vanadium metal. An open-top, cylindrical, mild steel vessel (150 mm diameter by 300 mm high) lined with alumina (50 mm thick) was the reduction reactor.

Oxygen is an unwanted, unavoidable but removable impurity in vanadium. Nitrogen is an also unwanted, avoidable but not easily removable impurity in vanadium. Any open (to air) process will result in nitrogen ingestion, and open aluminothermic reduction is not an exception. Nitrogen does not leave (degas from) vanadium during electron beam melting. Only fused salt electrorefining works. If the electrolysis process does not integrate well into the process flow sheet, it is best to avoid nitrogen pick-up in vanadium, which is possible. This is the reason for all commercial processes for vanadium using a closed bomb aluminothermic reduction. (Typical nitrogen pick-up: closed refractory-lined bomb reduction 0.006 wt.%, open refractory-lined bomb reduction 0.076 wt.%)

A significant development in carrying out aluminothermic reductions involves the use of a water-cooled copper crucible instead of a refractory-lined steel bomb, a technique pioneered by Perfect (1967, 1981), as already mentioned in Chapter 1. Such reactors enable reductions to be conducted under high-purity conditions (Carlson et al. 1981), because impurity pick-up from the refractory liner is avoided, and evacuation of the reactor volume and backfilling with pure inert gas is easier and effective to produce vanadium thermit under sterile conditions starting from very high-purity raw materials. The key impurities in thermit were (in wt.%): carbon 0.003, iron 0.006, nitrogen 0.002 and silicon 0.002. It has thus been possible to maintain a close control of the purity of the environment (crucible+cover gas) during these highly energetic reduction processes in a water-cooled copper reactor.

Many of the high-melting rare metals are produced relatively easily in the form of powder, even on an industrial scale, using an oxide or halide reduction by calcium, sodium, magnesium or hydrogen. These are relatively low-temperature processes and involve no liquid metal contact. The occasional use of refractory barriers between the reacting mass and the reactor material suffices to circumvent potential crucible problems (Jamrack 1963).

4.16   REACTIONS IN THE ABSENCE OF METAL VAPOUR

The reactions are carried out at relatively low temperatures, and no melt phases are involved. Crucible problems are usually not serious. A variety of refractory metals have been directly obtained in powder form (Hampel 1961; Jamrack 1963). For the oxide–calcium reaction for metal powder production, the reactants, oxide powder and calcium shreds are pressed together into pellets several inches in diameter and H/D=1. The pellets are stacked one on top of another, on a plate of sintered alumina or a thin layer of powdered lime inside a simple mild steel reactor, and reacted at 1000°C for long enough under argon.

For the production of thorium powder, kilogram batches of calcium mixed with thorium oxide loaded in nickel trays are passed through a furnace at 1000°C under argon flow. A steel crucible lined with molybdenum sheet can also be used for the reaction.

Sodium reduction of fluorides has been used to produce metal powders in unlined simple mild steel or stainless steel reactors. The powder product did not have any significant iron contamination. The charge, alternate layers of K_2TaF_7 crystals and sodium cubes, is stacked between deep layers of NaCl flux in a stainless steel bomb, which is heated externally to about 450°C to initiate the reaction.

An alternative method has also been used on a pilot scale for the production of capacitor-grade tantalum powder by sodium reduction of potassium tantalum fluoride (Jain et al. 1971; Sundaram et al. 1992), in Inconel-lined reactors at 925°C. Sodium is metered into a stirred melt of K_2TaF_7 at 925°C. From the reacted mass, tantalum powder is eventually separated.

Silica ware has been widely used in processes involving contact with metal chlorides as well as hydrogen chlorides and chlorine. Molybdenum trays loaded with thin layers of the relatively inert black solid $NbCl_3$ are heated to 800–1000°C in silica tube furnaces in hydrogen. In production facilities, multiple tubes each 3.5 m long by 150 mmdiameter, filled with molybdenum trays, are used to produce quintals of niobium metal powder. Molybdenum is very suitable for use in a chlorine environment as long as oxygen and water vapour are kept out.

4.17 KROLL REDUCTIONS

As preparation methods, heating a compound precursor in the vapour of a reducing agent or melting the metallic reducing agent in the vapour of the compound precursor often make their appearance, not only in exploratory experiments but also in industrial processes. Molten magnesium (b.p. 1091°C) reduces zirconium tetrachloride vapour in the well-known Kroll process. Lithium (b.p. 1330°C) and sodium (b.p. 883°C) are also used in chloride reductions (Jamrack 1963; Hampel 1961).

A metal is obtained in the form of a sponge when its compound (e.g. chloride) reacts with a molten metal reductant (e.g. magnesium or sodium) at a temperature (e.g. 850°C) sufficient to melt the resulting slag but not the metal itself. The reduced metal, initially formed in a finely divided form, coalesces in the molten slag and partially sinters into a sponge-textured material.

Magnesium chloride and sodium chloride are suitable slags due to their low melting points (712°C and 804°C, respectively) and low viscosity. Slag removal from the metal sponge is relatively easy – draining, distillation or leaching.

This method of metal making was originally developed by Kroll (1940, 1956, 1959) for titanium using titanium tetrachloride and magnesium. The technique has since been applied extensively to zirconium (Kroll et al. 1946, 1947, 1948, 1950)and also experimentally to vanadium and niobium (Hampel 1961; Jamrack 1963). This type of metallothermic reduction, the Kroll process, is perhaps the most important and also the most widely used. Metal–crucible interaction is generally not a major issue in Kroll reductions if temperatures are controlled, because molten metal–crucible contact does not happen. However, the reactive nature of chlorides results in the formation of the chlorides of certain impurity elements present in the crucible or reactor construction materials; these chlorides are co-reduced, and the impurity finds its way into the as-reduced metal sponge. It is impurity ingestion from the crucible to the reduced metal but happening through the vapour phase – more of a chemical transport process.

4.17.1 REDUCTION PLANT

The metal chloride reduction is always carried out in a closed reactor (Figure 4.12) in an argon atmosphere to keep out oxygen (and nitrogen). The reaction is exothermic, and it is necessary to limit the rate of reaction so that the temperature does not rise

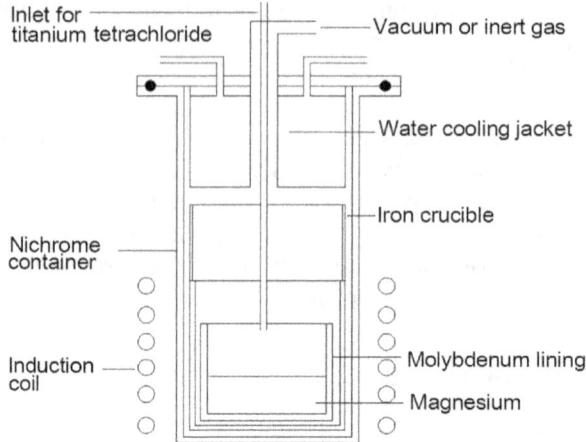

FIGURE 4.12   Titanium reduction furnace (From Darling, A.S. 1990).

to 1100°C (> the boiling point of magnesium) or fall too low, i.e. below the melting point of magnesium chloride (712°C). When a crust of MgCl$_2$forms, it blocks the chloride vapour from accessing magnesium. The temperature is, in practice, limited to between 720 and 920°C. Besides, alloys are formed between the zirconium or titanium (sponge) and the steel (reactor) at about 1000 to 1100°C, resulting in a fairly rapid penetration of the reactor or at least contamination of the sponge with iron. The reaction vessel is usually constructed of mild steel, 25/20 nickel/chromium steel or stainless steel except in the case of niobium production, because niobium penta-chloride is reduced by metallic iron even at quite low temperatures. A molybdenum liner is sometimes used as a protection from possible interaction (due to temperature excursions) between the steel reactor and the metal produced.

4.17.2   MAGNESIUM REDUCTION OF ZIRCONIUM TETRACHLORIDE

Magnesium reduction of zirconium tetrachloride was developed at the U.S. Bureau of Mines under the direction of W. J. Kroll. Compared with titanium reduction, the Kroll process for zirconium is procedurally elaborate. Zirconium chloride is sub-limed (it is a white solid that sublimes at 331°C) into molten magnesium, whereas liquid titanium tetrachloride was metered into molten magnesium. A suitable arrangement is incorporated for subliming zirconium tetrachloride in a controlled manner. Three heating zones are provided for in such furnaces.

4.17.3   BIMETAL REDUCTION

Reduction of ZrCl$_4$ using a mixture of sodium and magnesium metal is more advan-tageous than using only magnesium. The reduction is carried out initially with sodium at 600°C (10% of the ZrCl$_4$ is reduced with sodium) and then with Mg (for the remainder of ZrCl$_4$) at 800°C. Bimetal reduced sponge separated by pyro vacuum distillation is purest; the impurity concentrations (Cr,Al, Fe, Mg) are lower. The reasons are the gettering ability of sodium, lower solubility of Fe in Na (Fe from the reactor/crucible), better fluxing ability and hence, scavenging action of NaCl with impurity chlorides ($FeCl_3$, $AlCl_3$) and better separation of $MgCl_2$–NaCl slag (more fluid than $MgCl_2$ alone). NaCl is a better conductor of heat, and temperature excursions in the reactor are reduced, resulting in purer sponge.

4.17.4   CRUCIBLE CONTAMINATION

The purity of sponge is also dependent on its location in the reduction crucible. Ti sponge at the bottom of the crucible is less pure compared with that in the middle of the crucible. The metal in the top periphery is fine and pyrophoric. The reasons for variations in purity and form are variations in local reaction conditions and also extent of contact with the crucible.

Vanadium sponge was produced by the Magnesium Electron Company, United Kingdom, at a 20 kg batch size by magnesium reduction of vanadium trichloride in a mildsteel reactor. After reduction and distillation, vanadium was obtained as a single piece of fairly soft sponge. Vanadium from near the crucible wall has a higher iron content (Jamrack 1963).

4.17.5   MAGNESIUM REDUCTION OF NIOBIUM AND TANTALUM PENTACHLORIDES

While the steel reactors worked well enough for titanium, zirconium and vanadium in implementing the Kroll-type process for metal sponge production by chloride reduction, vapour-phase corrosion of the steel reactor derailed the process in the case of niobium and tantalum (Jamrack 1963).

Niobium pentachloride vapour attacked the metallic iron of the reactor, forming niobium trichloride on the reactor wall and releasing ferric chloride into the vapour phase. On contacting the molten magnesium, ferric chloride was reduced to iron, which entered the product. Nickel-plating of the steel reactor did reduce the iron content in niobium, but a reactor made entirely of nickel would solve the problem of iron contamination. The results of a separate investigation on tantalum pentachloride were similar.

4.18   LITHIUM REDUCTION OF RARE EARTH TRICHLORIDES

The rare earth metals are, as a group, extremely reactive and have the reputation of being difficult to prepare. The reactivity of rare earth metals really shows up at high temperatures and when they are molten.

Nolting et al. (1960) and Moriarty (1968) used reduction of rare earth trichloride by lithium vapour to prepare yttrium and other high-melting rare earth metals. Nolting et al. (1960) carried out the reductions at 950°C in a stainless steel reactor lined with molybdenum foil (Figure 4.13). The molten rare earth chloride was contained in a tantalum crucible and held at 950°C for 2 h in the reactor filled with lithium vapour. The vapour reduced yttrium trichloride to yield yttrium crystals and

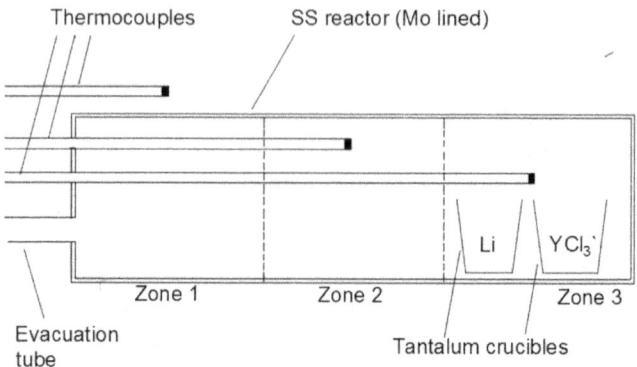

FIGURE 4.13   Reduction bomb assembly for lithium vapour reduction (From Nolting, H.J., Simmons, C.R., and Klingenberg, J.J. 1960).

lithium chloride slag, which were eventually separated. Yttrium was 99.8% pure with 0.16% oxygen as the major residual impurity. The as-reduced metal was in contact at any time only with tantalum, and there was no contamination from tantalum. Metal–crucible interaction is generally circumvented by preparing the rare earth metal without melting it.

Lithium reduction of rare earth trichlorides was carried outon a commercial scale by Moriarty (1968). Reduction was carried out at by holding molten rare earth chloride in titanium crucibles at 800–1000°C in a stainless steel chamber filled with lithium vapour. The crucible arrangement was in principle the same as that used by Nolting et al. (1960). The lithium reacted with the molten rare earth chloride, reducing it to rare earth metal and forming lithium chloride slag. Moriarty (1968) reported the preparation of all the rare earth metals except promethium, samarium, europium and ytterbium. Apart from stating that the contamination of the reduced rare earth metal crystal with the crucible material was very low, no values were given by Moriarty. The major impurity in these metals was oxygen, which ranged from 225 ppm for holmium to 1600 ppm for terbium.

In a significant effort to reduce residual oxygen in the as-reduced rare earth metal, Croat (1969) used lithium–5% calcium as reductant in place of pure lithium. Any oxygen in the lithium vapour was gettered by calcium. The reduction was entirely carried out in tantalum crucibles inside a tantalum retort. The as-reduced metals were of 99.94 wt.% or 99.2 atomic% purity. The tantalum contentswere below 30 ppm. The investigation by Croat (1969) actually served to define the limits to which purity levels of as-reduced rare earth metals can be raised in a real process.

In the same year as Nolting et al. (1960) published their work on lithium vapour reduction of yttrium trichloride, Block et al. (1960) reported on the reduction of yttrium trichloride, but with sodium instead of lithium. The arrangement and procedure were very similar to the Kroll process for zirconium. Sodium and yttrium chloride were loaded into two separate molybdenum containers in the stainless steel retort. Sodium was vaporized and the sodium vapour reduced yttrium chloride. The reduction temperature was 850°C. Molybdenum contamination in the yttrium was lower than 100 ppm in both cases. However iron, chromium and nickel contents were not insignificant, some probably being derived from the reactor.

4.19   INGOT REDUCTION PROCESSES

The classic closed bomb metallothermic reduction method that was used for producing ingots of uranium, thorium and vanadium metals was also used for the rare earths. The reactors, charge composition and general reduction procedures were similar. A combination of rare earth chlorides and calcium was used. Rare earth chloride reductions are not known for exothermicity. A booster reaction, usually Ca–I, was always used. The oxide liner, sintered lime or dolomitic oxide, was sufficiently inert to molten rare earth metals Ce, La, Pr and Nd only up to about 1020°C (Spedding and Daane 1961;. Beaudry and Gschneidner Jr 1978).

4.19.1   REDUCTION IN A TANTALUM CRUCIBLE

Two major problems confronted the attempts made to prepare high-melting rare earth metals by chloride reduction. At high temperatures, the calcium chloride frothed and prevented clean slag–metal separation, and the oxide liners were attacked by the metal and impurities were introduced. Subsequently, Spedding and Daane (1952) attempted the reduction in a tantalum crucible. The procedure was as follows.

In a typical run, rare earth chloride and calcium were mixed and either jolt packed into a tantalum crucible or compacted in a powder press into cylinders that were then placed in the tantalum crucible. The loaded crucible was covered with a perforated tantalum lid and heated in an induction furnace under argon to 1000°C for lanthanum, cerium, praseodymium and neodymium and 1350°C for gadolinium. The melt was held for 13 min to permit complete coalescence of the product metal. After cooling to room temperature, metal and slag were separated. The use of a booster reaction was not necessary, because external heating was used. Tantalum was more resistant to molten rare earth metal attack than any refractory material.

The method of producing pure rare earth metals by the chloride reduction route, though feasible for most of the metals, is usually an elaborate procedure requiringgreat procedural care and experimental facilities, particularly on account of their hygroscopicity and volatility. These two limitations are, to a large extent, circumvented by the use of fluorides in place of chlorides.

4.19.2   REDUCTION OF RARE EARTH FLUORIDES

The process was developed at Ames. Calcium was used as the reducing agent. The lower vapour pressure of calcium fluoride permitted reduction temperatures as high as 1700°C, higher than the melting point of any rare earth metal. It was thus possible to prepare all the rare earth metals (except those exhibiting stable divalency) irrespective of their melting point.

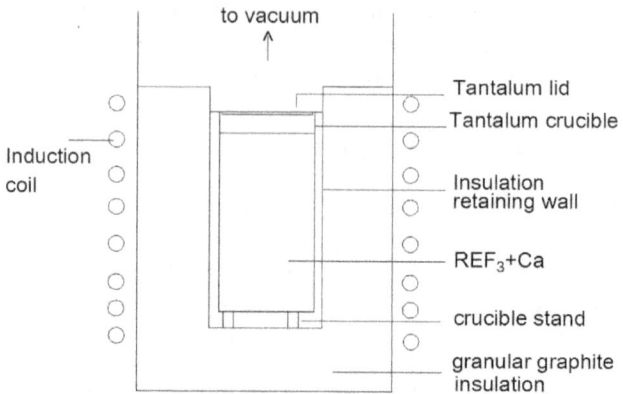

FIGURE 4.14   Tantalum crucible for rare earth metal production (From Daane, A.H. 1961a).

4.19.3   CALCIUM REDUCTION (AMES PROCESS)

In the Ames process (Daane and Spedding 1953), the reductions were carried out by heating the mixture of calcium and rare earth fluorides in a tantalum crucible in argon in an induction furnace (Figure 4.14. The reaction initiated at 800–1000°C and was not violent. Heating was continued to a temperature where both the products, slag and metal, were molten, i.e. 50°C above the melting point of calcium fluoride or 50°C above the melting point of the metal being prepared, whichever was higher. Once the furnace had cooled to room temperature, the crucible was removed from the furnace in air, and the slag layer that had formed on top of the metal ingot was knocked out of the crucible, leaving a clean ingot of rare earth metal in 97–99% yield with 0.1–2% calcium as the principal impurity. Sufficient thickness of the tantalum crucible permits its repeated use (Daane 1961b). Yttrium is the highest-melting rare earth. Spedding et al. (1970) gave a nice picture, reproduced here as Figure 4.15, showing the cross section of a typical reduction crucible in calciothermic reduction of rare earth trifluorides. The sketch clearly shows the distinctly concave top of solidified rare earth metal and the convex top of the solidified CaF_2 layer. Molten rare earth metals wet tantalum, and molten fluorite does not wet tantalum.

4.19.4   GOLDSCHMIDT PROCESS

In the industrial practice described by Herget (1985) for calcium reduction of rare earth fluoride, the fluoride was mixed with granular calcium, and the mixture was placed in a tablet form or as rammed material in a steel reactor. The reactor was lined on the inside with tantalum, which was coated with a layer of calcium fluoride. After the reactor was closed and evacuated, the metallothermic reduction was initiated by a spark discharge. The reduction proceeded to completion within minutes.

$$2REF_3 + 3Ca = 2RE + 3CaF_2 \tag{4.5}$$

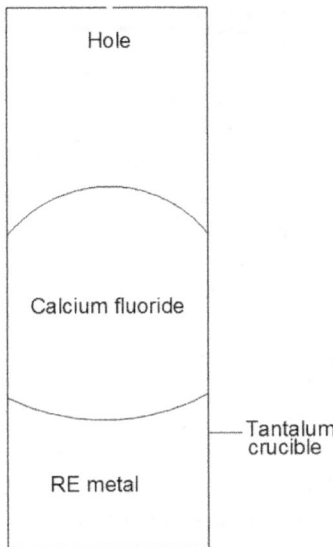

FIGURE 4.15 Crucible cross section after calciothermic reduction of rare earth fluoride (From Spedding et al. 1970).

The rare earth metal collected as an ingot in the bottom part of the reactor, topped by solidified calcium fluoride slag.

The preparation of rare earth metals by reduction of their trifluorides with calcium in tantalum crucibles leads to the introduction of up to 0.5% tantalum as impurity. This happens in particular in higher-melting rare earths, because higher temperatures are reached in their production, and the solubility of tantalum in rare earths increases with temperature (Dennison et al. 1966a, 1966b). In the case of scandium, the tantalum content goes up to 2–5%.

The temperature necessary for fluoride reduction was lowered, and thus, the contamination of crucible material in the resulting rare earth metal was decreased accordingly through variants of the basic process. The process now proceeds through the formation of an intermediate low-melting alloy with zinc (Spedding et al. 1960).

$$2ScF_3 + 3Ca + 3Zn + LiF \rightarrow Sc - 60\% \; Zn \; alloy$$

$$+ 3CaF_2 - LiF\left(eutectic \; composition\right) \quad (4.6)$$

The alloying of scandium (m.p. 1541°C) and formation of the low-melting slag made it possible to carry out the reduction and separation of the products into molten metal and slag layers at 1100°C. The reaction is carried out in a tantalum crucible. After reduction, the product:scandium–zinc alloy, which also contained 1–2% Ca, was crushed and vacuum distilled to remove zinc, leaving behind pure scandium as a porous sponge containing only traces of tantalum. This process is similar to the calcium reduction process for the preparation of thorium, wherein a thorium–zinc alloy is prepared and eventually converted to pure thorium metal (by distilling off zinc).

4.20   OXIDE REDUCTION PROCESSES

Rare earths form highly stable oxides. They cannot be reduced to metal by an oxide reduction process in the classical sense. However, not all rare earth metals can be produced by halide reduction either. Samarium, europium and ytterbium exhibit stable divalency in addition to trivalency. On reduction, the trivalent halide is reduced to the divalent halide, and the process stops there.

In the 1950s, Daane, Dennison and Spedding (1953) at Ames devised the reduction–distillation process. This method, which is similar to the Pidgeon process for magnesium (Pidgeon 1944), has been remarkably useful for the preparation of samarium, europium and ytterbium.

Lanthanum is the least volatile of the rare earth metals; dysprosium has a vapour pressure nearly 300 times that of lanthanum at the same temperature. The metals samarium and europium are similarly very volatile. Besides, lanthanum oxide has the most negative heat of formation among the rare earth oxides. These observations led Daane et al. (1953) to devise a method for reacting the oxides of samarium, europium or ytterbium with lanthanum metal and driving this reaction to completion by distilling away in vacuum the volatile metals formed as product. The free energy of the reaction was made negative by strongly influencing the value of Q (Chapter 2). The method has since been called lanthanothermic reduction or lanthanothermy.

An important feature of the reduction–distillation process is that it is forced to quantitatively yield the product required by removal (by vaporization) of the rare earth metal formed from the reaction mixture. The reactor and crucible material used is tantalum. Maximum reduction temperatures range from 1450 to 1600°C, and the metals obtained are usually very pure with respect to lanthanum and tantalum. Interaction with the crucible has not developed into an issue. Significantly, no melts are involved. In fact, a Goldschmidt-type reduction of rare earth oxides to yield rare earth metal ingot has strong heat of reaction and slag melting barriers.

4.21   FUSED SALT ELECTROLYSIS

Rare metal electrolytic processes are invariably carried out in an inert atmosphere. The rare earth metals that have been produced by fused salt electrolysis of chlorides on a commercial scale (Morrice and Knickerbocker 1961; Hirschhorn 1968) are lanthanum, cerium and didymium (Nd + Pr). The cells had an iron-, carbon-, graphite- or refractory-lined steel vessel to contain the molten bath. The cathode was the bath container itself or an iron or carbon block at the bottom of the container. The anode was a set of graphite rods lowered into the cell through its top. Besides $RECl_3$, the electrolyte contained NaCl, KCl and $CaCl_2$, and the bath was molten even at 650°C.

Iron contamination was greater for cerium (0.19%) than for lanthanum (0.11%) or didymium (0.05%), even though the electrolysis temperature was the lowest for cerium. The maximum operating temperature of the chloride cell is 1100°C, and neodymium (m.p. 1021°C) is the highest-melting rare earth metal that can be produced by chloride electrolysis, albeit in low yield. At high temperatures in a chloride electrolyte bath, the volatility of bath constituents, the solubility of metals in their chlorides, and the attack of cell wall materials by the rare earth metals become severe. Considerably higher-temperature operation, however, is possible in cells using fluoride electrolyte baths.

4.22 OXIDE–FLUORIDE ELECTROLYSIS

4.22.1 GRAY'S CELL

The most successful among the early experiments on oxide–fluoride electrolysis was reported by Gray (1951). CeO_2 was dissolved in a CeF_3–LiF–BaF_2 electrolyte contained in an externally heated carbon container with 57 mm inside diameter. The anode was of high-purity graphite rod, and the cathode was a molybdenum rod; both were suspended from the top and immersed in the melt. A molybdenum crucible was kept below the cathode to collect the product. Unless it is oxidized, molybdenum does not dissolve in cerium. The bottom of the cell was lined with a thin sheet of molybdenum to prevent any accidentally spilled cerium reacting with graphite to form cerium carbide. Molten cerium metal dropped off the cathode and collected in the molybdenum cup kept below. The overall purity of the metal was 99.7–99.8% cerium. The product contained 0.1–0.6% calcium and up to 0.3% magnesium as major impurities.

Many types of electrolytic cells were designed and used at the Reno facility of the US Bureau of Mines for carrying out oxide–fluoride electrolysis (Morrice and Wong 1979). The first was a 150-mm (6-inch) diameter electrowinning cell, known as cell Type 6 (Morrice and Knickerbocker 1961; Morrice et al 1961).

4.22.2 RENO CELL TYPE 6

The cell consisted of a 150-mm (6-inch) diameter bath container made of graphite. Three 10-mm diameter molybdenum cathodes and three 25-mm diameter carbon anodes extended down into the container through its top. This electrode arrangement was first used to melt the bath with alternating current and later used for electrolysis with direct current.

The cell was arranged to operate fully under an argon atmosphere. It was internally heated both for raising the temperature from cold and later, for sustaining the temperature sufficient for electrolysis, 810–830°C. One of the important features of the cell was the formation of a layer of frozen bath or skull on the interior surface of the graphite container bottom. The skull prevents the molten electrowon rare earth metals from contacting the graphite crucible. The thickness of the insulation on the side walls and bottom of the graphite cell, and the geometry of the electrode with respect to the side wall and bottom of the cell, were chosen so as to maintain this electrolyte skull.

4.22.3 RENO CELL TYPE 12

These were the electrolytic cells in which the electrolyte was contained in a ~300-mm (12-inch) diameter graphite crucible. The key feature of the cell was that the bottom and sides of the graphite crucible were lined with a molybdenum sheet and a tapping pipe was attached to the molybdenum hearth so that the molten metal could be tapped. The tapping pipe extended into an inert atmosphere casting chamber.

A typical continuous electrowinning run extended over 3 days, during which metal was electrowon and several tappings of homogeneous cerium metal were made. The metal purity was 99.8%. Molybdenum figured as a major impurity. It was picked up by cerium during the extended contact with the molybdenum hearth. Tungsten contamination under similar circumstances would be much less.

4.22.4   HIGH-TEMPERATURE ELECTROWINNING CELL

A high-temperature electrowinning cell, also from Reno (Henrie and Morris 1966), operating in the temperature range 1370–1700°C and used for the preparation of liquid gadolinium, dysprosium and yttrium, is shown in Figure 4.16.

The cell was heated internally by DC, and the electrolysis current was supplemented by AC applied between the graphite anodes, which supplied approximately 75% of the total power. The cell bottom could be cooled by passing helium or air through the copper coil placed below the graphite crucible to create an electrolyte skull for collecting the metal. The cell, like all the others from Reno, was enclosed in an inert atmosphere (controlled atmosphere temperature and pressure) chamber to prevent reaction of the cell components and chemicals with the atmosphere.

In the construction of the high-temperature cell, the materials were taxed to the extreme, and Morrice et al. (1968) used a combination of materials, as shown, to design the cell. The electrolyte for gadolinium electrowinning consisted of an equimolar mixture of GdF_3 and LiF. Gadolinium oxide was fed to the molten electrolyte maintained at 1370°C. The metal was deposited as liquid at the cathode and dropped down to collect as nodules in the frozen electrolyte skull at the bottom of the graphite crucible. The temperature here was maintained at 810°C by cooling the bottom of

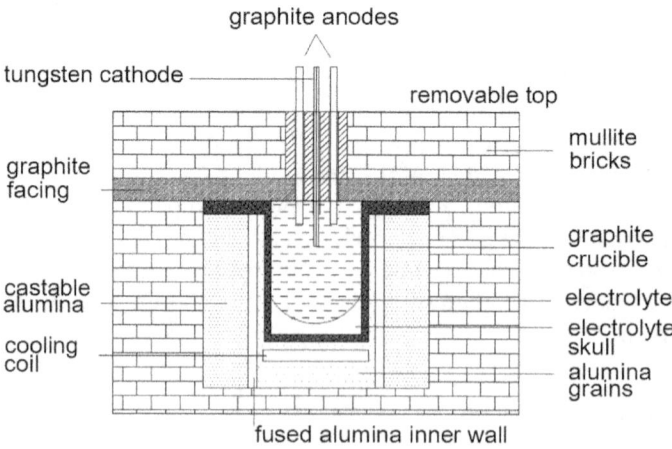

FIGURE 4.16   High-temperature electrolysis cell for rare earth metals (From Morrice et al. 1968).

the cell. Such a difference between the temperatures of deposition and collection was essential to achieve high recovery and to ensure high purity of the metal product. A typical gadolinium nodule contained 0.05% carbon, 0.01% oxygen and 0.05% tungsten as major impurities. Liquid yttrium and dysprosium were also electrowon from their oxides dissolved in melts composed of 50 mole% LiF and 50 mole% YF_3 or DyF_3. The metals were deposited on the cathode at 100 to 200°C above their melting points (m.p. of Dy 1412°C, m.p. of Y 1522°C) and collected at temperatures 400 to 500°C lower. Yttrium and dysprosium nodules thus recovered contained as little as 0.08% carbon and 0.10% oxygen. The highest temperature at which the cell was operated was 1700°C, for electrowinning yttrium.

4.23   CRUCIBLE CONTAMINATION IN RARE EARTH REDUCTION

A reduction carried out in a tantalum crucible results in an average of 0.05% tantalum in the as-reduced metal (Huffine and Williams 1961). This concentration can be much higher if reduction is carried out at higher temperatures or for longer periods of time. A tantalum crucible was used by Croat (1969) in the preparation of Dy, Ho and Er. These metals were prepared by the reduction of their anhydrous chlorides with lithium. The temperature of reduction was 900°C, which was well below the melting point of these metals. The tantalum contamination of as-reduced Dy, Ho and Er was quite low, i.e. <30 ppm. This can be contrasted with the tantalum content of 0.1–0.2% in the as-reduced form of Dy, Ho and Er when they are obtained by calcium reduction of their fluorides in tantalum crucibles. These reductions are carried out at high temperatures, above the melting point of the rare earth metals.

Daane (1961a) stated that because the solubility of tantalum in liquid rare earth metals increases measurably with increasing temperature, the lowest possible temperature is used in the processes wherein the rare earth metal is processed in contact with tantalum in order to keep the tantalum impurity as low as possible. In the reduction of the light rare earth metals (lanthanum through neodymium), the amount of tantalum in the metal is approximately 200–300 ppm. For higher-melting heavy rare earths and yttrium, this figure increases to about 1000–5000 ppm. Among the rare earth metals, tantalum is the most soluble in scandium. As much as 3% tantalum is present in scandium melted in tantalum. Tungsten is better as a crucible material. In the liquid rare earth metals, less tungsten than tantalum dissolves at any temperature (Dennison et al. 1966a, 1966b). The refractory metals molybdenum, niobium, titanium and zirconium are distinctly more soluble in molten rare earth metals than are tantalum and tungsten.

In reduction procedures, where an alloying element is used to lower the melting point of the as-reduced metal (for example magnesium in yttrium reduction), titanium or zirconium may be used as crucibles. When yttrium–magnesium alloy was reduced in titanium crucibles, the alloy picked up an average of 0.15% titanium, and when zirconium was used, the pick-up was 0.58% zirconium (Huffine and Williams 1961).

In the electrolytic preparation of the rare earth metals, cell corrosion can occur extensively, and the corrosion products can end up in the metal produced. Carbon impurity in the electrolytically produced metals usually comes from graphite anode and/or cell walls (Beaudry and Gschneidner 1978). Besides carbon, depending on the materials used for cell construction, the metals may be contaminated with iron, silicon and other metallic impurities. The use of Mo as a liner or crucible in the cell leads to pick-up of the element in the electrolytic metal. Electrolytic cerium metal obtained using Mo as cell material showed a Mo level of >750 ppm. Tantalum was not picked up significantly (Shedd et al. 1964) when used in place of molybdenum.

4.24 PYROVACUUM TREATMENTS

4.24.1 RARE EARTH METALS

The rare earth metals can be divided into four groups primarily on the basis of their volatility. In the post-reduction vacuum treatment, each group is processed somewhat differently (Beaudry and Gschneidner 1978).

4.24.1.1 Lanthanum, Cerium, Praseodymium and Neodymium

These four light rare earths have relatively low melting points, very high boiling points and hence, extremely low vapour pressures at the melting point. High-temperature vacuum melting is the only process by which they are purified.

As-reduced La, Ce, Pr and Nd from the fluoride process just separated from the CaF_2 slag contain Ca, CaF_2 and F, and H. All these impurities are more volatile than these rare earth metals and are removed by induction melting the metals in a vacuum for 10 min. Melting of La and Ce was carried out at 1850°C; for Pr, the temperature was 1750°C, and for Nd, it was 1650°C (Beaudry and Gschneidner 1978). A tantalum crucible was used. At the high temperatures used for vacuum melting, the rare earth metals dissolve a considerable amount of tantalum (Dennison et al. 1966a, 1966b). However, the solubility of tantalum in these metals decreases with temperature, and it is very low at the metals' melting point. This information was utilized in restricting the amount of tantalum impurity in the rare earth metal.

After vacuum melting the metals at high temperature, the furnace was slowly cooled to about 10°C above the melting point of the metal to allow the Ta to precipitate out of the solution and settle to the bottom of the crucible. Later, when the ingot was removed from the furnace, the thin tantalum crucible and the dendrites that had settled to the bottom of the crucible were machined off.

4.24.1.2 Yttrium, Gadolinium, Terbium and Lutetium

The complete pyrovacuum purification sequence for these metals consists of vacuum melting followed by distillation. As in the case of the first group of metals (La, Ce, Pr and Nd), the use of a tantalum crucible in a vacuum melting at ~1800°C leads to the dissolution of a certain amount of tantalum in Y, Gd, Tb and Lu. Unlike in the four light rare earths, however, the tantalum solubility in Y, Gd, Tb and Lu does not decrease much on reducing the temperature to their melting points, and therefore, the tantalum in the solution does not precipitate out by cooling. The only way to remove tantalum is by distilling off the rare earth metals, leaving tantalum in the residue. If tungsten crucibles were used instead of tantalum in reduction and vacuum melting, tungsten would dissolve in the rare earth. But after vacuum melting in these metals, tungsten could be precipitated out in exactly the same way as Ta was in La, Ce, Pr and Nd. This alternative is available if the 0.012 at .% W that will persist in the Gd after precipitation can be tolerated. Similarly, at the end of precipitation, 0.03 at .% W will remain in Tb and 0.07 at.% W in yttrium (Dennison et al. 1966a; Beaudry and Gschneidner 1978).

The distillation of Y, Gd, Tb and Lu was carried out at Ames in tungsten crucibles, because refluxing of some of the distilled metals eroded the tantalum crucibles, causing them to leak. A tungsten-lined condenser was used, since tantalum diffuses into the hot condensate more readily than tungsten.

4.24.1.3 Scandium, Dysprosium, Holmium, Erbium and Lutetium

These metals have appreciable vapour pressure at their melting points. The vacuum melting, therefore, had to be relatively brief, or much of the metal would be lost by vaporization. The other part of pyrovacuum treatment in the refining of Sc, Dy, Ho and Er is sublimation. The high volatility of these metals is an advantage here, and purification of metals by sublimation is comparatively easy. Tantalum introduced into the metals during the vacuum melting step was removed by subliming off the metals, leaving tantalum in the residue. Lutetium has been purified by distillation as well as by sublimation (Spedding et al. 1968). When distillation was carried out at 1850°C in tantalum crucibles, tantalum dissolved in the refluxing liquid lutetium, and the liquid metal often eroded a hole through the tantalum crucible. However, the use of a tungsten crucible and tungsten-lined condenser enabled distillation of lutetium without extensive crucible attack (Beaudry and Gschneidner 1978).

4.24.1.4 Samarium, Europium, Thulium and Ytterbium

These four metals are prepared by the lanthanothermic or reduction–distillation method. Tantalum does not become the major impurity with the metals Sm, Eu, Tm and Yb, because they do not undergo the high-temperature vacuum melting step. Hence, there is no crucible interaction issue.

4.24.2 Uranium

The results of uranium melting in graphite, tantalum, yttria and yttria-coated crucibles are summarized in Table 4.2. Yttria coating by plasma spray (Chakravarthy et al. 2015) not only effectively isolates the crucible proper from molten uranium but also does not itself contaminate uranium or become degraded by it.

4.24.3 Vanadium

Vanadium is a relatively difficult metal to prepare (Gupta and Krishnamurthy 1992). It is likewise a difficult metal to purify. Vanadium canbe produced by widely different methods from a variety of its compounds (Rostoker 1958). In every case, there are issues relating to containment of vanadium, especially when a vanadium melt contact with the container is involved. There are examples of problems and solutions

TABLE 4.2
Crucibles for Uranium Melting

Crucible (coating if any)	Melting conditions: temperature,°C (time, min), atmosphere	Results	Reference
Graphite (yttria coating)	1300–1500 (30)	No carbon pick-up	Koger et al. 1976
Yttria – polycrystalline, sintered	1400 (12,000)	Uranium picked up oxygen from yttria, $UO_2 \cdot Y_2O_3$ formed at the melt–crucible interface Solubility of Y in U is 0.5 at .% at 1300°C	Tournier et al. 1998
Tantalum	Molten uranium	Crucible punctured by intergranular attack	Kuznietz et al. 1988
Tantalum (yttria coating 80–100 μm by thermal spray)	1300 (4800)	No attack or impurity pick-up or interdiffusion of U into yttria	Alangi et al. 2011
Graphite (yttria coating 300 μm by atmospheric plasma spray)	1200 (120)	Yttria effectively isolated uranium melt from graphite	Chaktavarty et al. 2015
Yttria	1600 (15)	U-10 wt.% Zr alloy was melted without interaction	Kim et al. 2013

when vanadium is produced by metallothermic reduction of oxides and also when the metal is obtained by metallothermic reduction of its chloride.

As-reduced vanadium is usually obtained in 80–99%purity. It contains many metallic and non-metallic impurities. Among the metallic impurities, iron is usually present along with aluminium, calcium, chromium, copperand occasionally magnesium. The non-metallic impurities invariably present are carbon, oxygen, nitrogen and silicon. Certain impurities in the metal can be avoided or decreased by a suitable choice of reduction conditions and by using high-purity start materials. However, it has not been possible to obtain, in the reduction stage itself, vanadium metal pure enough with respect to all the impurities. Post-reduction purification is an unavoidable step in pure vanadium production.

Vacuum induction melting (VIM) is not only great as a refining process but also integrates very well with many good fabrication techniques. The major problem with the VIM method for consolidation and purification of vanadium concerns the container material. Molten vanadium attacks most, if not all, the usual refractory materials and becomes impure because it picks up oxygen, carbon or other impurities from the crucible. The crucible is also rendered useless because liquid vanadium seeps through it. Rostoker (1958) has summarized the results of melting vanadium in some of the refractory crucibles as in Table 4.3. Interestingly, the list could not be

TABLE 4.3
Results on Induction Melting of Vanadium in Various Crucibles

Crucible material	Time held molten, min	Weight melted, g	Hardness of vanadium as cast, Vickers Hardness Number	Crucible attack	Remarks Impurity pickup and product condition
Recrystallized alumina	0.75	20	410	Slight	
Al_2O_3			569		1.52% O pick-up, 0.20% N pick-up
Alundum	0.5	26.6		Very severe	No button
BeO	0.5	20.3	221	Very slight	
BeO	2	35	251	Slight	
BeO			564		1.15% O pick-up, 0.21% N pick-up
CaO	5	30	423	Very severe	
CaO			545		1.13% O pick-up, 0.13% N pick-up
CeS			182		0.04% O pick-up, <0.04% N pick-up, 0.15% S pick-up
Graphite (grade ATZ)	4	50	325	Slight	2.98% C pick-up
Graphite (grade AUC)	0.75	50	315	Slight	3.91% C pick-up
Graphite (grade AUC)	8	200	280	Slight	3.29% C pick-up
Graphite (grade AUC)	8	500	298	Slight	2.08% C pick-up
MgO		40.2		Very severe	No button recovered
Magnorite	0.5	26.6	564	Very severe	
Carbon-bonded SiC	30			Very severe	No button recovered
ThO_2	1	27	247	Slight	
TiO_2				Very severe	
ZrO_2 (stabilized)	1	39	438	Severe	

Source: Rostoker 1958, Merrill 1958.
General:
Atmosphere: helium
Hardness of vanadium as arc melted: 217 VPN
Analysis: 0.05% O and <0.04% N

updated with new additions in the last six decades, because there have been none. Fabrication of vanadium alloys have been by the routes involving vacuum arc melting or electron beam melting (Gupta and Krishnamurthy 1992).

Among the oxide crucibles used, beryllia and thoria withstood the attack of molten vanadium. This result is applicable only for very short melting times (0.5 to 2 minutes). The pick-up of oxygen by molten vanadium from the oxide crucibles is very high and sufficient to render the metal useless for many applications. Liquid vanadium only slightly attacks graphite crucible for upto about 8 minutes of melting time but picks up between 2% and 4% carbon in the process. The amount of carbon picked up by the molten vanadium strongly depends on the quantity of vanadium melted in a given crucible. This points to the possibility that melt–crucible contact area is a major factor determining the ingestion of carbon, more even than the duration of melting. The carbon pick-up from graphite by different quantities of vanadium melted shows this. A similar situation will be encountered later in the melting of NiTi alloys in graphite (Otubo et al. 2006), where the batch size of the melt plays a decisive role in contamination. This strong carbon contamination is not reflected in the hardness values of as-cast vanadium, because carbon solubility in vanadium is limited, and compared with other interstitials, the contribution of carbon to the increase in the hardness of vanadium is less. Among the other refractories tested, cerium sulphide (CeS) was found to be good in that it did not contaminate liquid vanadium with oxygen. However, some amount of sulphur (0.15%) got into the metal (Rostoker 1958).

Regarding the refractory metals tantalum, molybdenum and tungsten, tantalum and tungsten cannot be used to hold molten vanadium, because even before vanadium melts, these metals form a liquid phase with vanadium due to the occurrence of a minima in the liquidus at a composition near pure vanadium. Even though no such minima occurs in the Mo–V system, these metals form a continuous series of solid and liquid solutions over the entire composition range. It therefore appears that no crucible – metallic or ceramic – can contain liquid vanadium without being attacked and/or contaminating vanadium. To hold molten vanadium for a very short time (2 min), beryllia and thoria crucibles appear satisfactory. Rostoker (1958) has indicated the possibility of using beryllia- or thoria-coated crucibles to melt vanadium. However, none of the crucibles described earlier can be used to melt vanadium for refining the metal, because it will be necessary not only to superheat the liquid metal but also to contain the vanadium melt for a longer time to facilitate complete refining. Thus,VIM does not appear to be a workable melt refining process for vanadium.

According to the Ellingham diagram, the oxides MgO, CaO, ThO$_2$ and BeO are all far more stable than the lowest (and most stable) vanadium oxide, VO(s), at all temperatures. That beryllia and thoria are resistant to attack while magnesia and calcia are strongly attacked by molten vanadium suggests that factors other than simple thermodynamic stability of oxides may be at play. The literature information available so far on molten vanadium–crucible interaction does not provide details on the wetting behaviour and contact angle data, density and porosity conditions of the crucibles tested, or the interface between the melt and the crucible. One or more of these may help in reconciling observations with known thermodynamic and kinetic behaviour and provide clues to more resistant crucibles for use in VIM.

4.25 TITANIUM

The industrial importance and the quantity of research effort that titanium has attracted in the past 80 years among the less common metals is comparable to those of iron and steel among the common metals. Titanium is a very abundant element in the earth's crust, and its major ores are easily accessible in every part of the world. The question of titanium metal production in sufficient purity and decent quantities was presented with a workable solution when William Justin Kroll developed a magnesium reduction process for titanium tetrachloride, the 'Kroll process', in the 1940s (Kroll 1940, 1959). Certain titanium alloys have outstanding combinations of properties that make them highly attractive for applications as structural materials in the aerospace and automotive industries, for chemical, petrochemical and marine applications, and in the manufacture of biomedical components and surgical instruments. Some other alloys of titanium have a combination of properties – shape memory and hydrogen storage – that make them very useful as functional materials for special applications in power generation, energy storage, robotics, biomedical and telecommunications, among others.

TiAl-based alloys are difficult to prepare and have low ductility at room temperature and low formability. Precision casting is the way to go for these alloys (Lapin 2009; Freuh et al. 1996, 1997). The critical process is alloy melting. The crucible–melt interactions in a VIM operation depend on the titanium content of the alloy as well as the temperature needed for its melting. In practice, TiAl(γ) alloys can be melted at temperatures between 1550 and 1750°C, while Ti6Al4V alloys require 1660–1800°C due to the high Ti content (Fashu et al. 2020). Until now, no refractory material has been found to be absolutely inert against these alloys, and some interactions between melt and crucible and even mould materials have always occurred during melting and casting (Fashu et al. 2020; Freuh et al. 1996, 1997; Uwanyuze et al. 2021). These melt–ceramic material interactions lead to contamination and degradation of properties (Tetsui et al. 2012a, 2012b). If titanium and its alloys can be melted and cast without significantly degrading the crucible and/or the metal itself, in sufficient volumes and inexpensively, it will be considered, and remain for eternity, a major milestone in the annals of materials processing.

4.25.1 MELTING TECHNIQUES

Titanium and its alloys tend to dissolve substantial quantities of impurity elements. Hydrogen can be removed by vacuum treatment, but oxygen and nitrogen, once dissolved, cannot be readily removed by any simple or straight forward method. The focus is therefore on not allowing the metal to ingest the impurities rather than trying to remove them once ingested.

There are different techniques to manufacture titanium alloy parts. Investment casting is one of them and probably, the most significant. Investment casting is an established process for manufacturing near-net-shape intricate parts. The technique is known and has been practised for centuries as lost wax casting. It is carried out in ceramic shell moulds and is preceded by melting of the alloy. However, due to the low fluidity of titanium alloys, the melt needs to be taken to a high superheat. Presently used furnaces for melting titanium alloys typically have water-cooled copper crucibles and an inert gas cover. Temperatures sufficient for melting, however high they may be, are attained easily using arc, electron beam and also induction heating. But after melting occurs, raising the temperature of the melt much further is either not possible (vacuum arc remelting [VAR]) or beset with cost problems (induction skull melting [ISM], electron beam melting [EBM]). These melting techniques and equipment – ISM (Harding et al. 2011; Harding and Wickins 2003), VAR and EBM (Ou et al. 2015; Sakamoto et al. 1992) are specifically made for alloys that are very reactive in the molten state. They are, by nature, exceedingly energy intensive, accommodate only a small metal volume and are only able to achieve a small superheating temperature (Harding and Wickins 2003). VIM in a ceramic crucible is significantly more energy efficient and capable of much higher superheating and thus, is more conducive to obtaining a heat-laden melt pool suitable for larger castings (Fadeev et al. 2015; Kamyshnykova and Lapin 2018; Gomes et al. 2011). Thus, the key requirements for a durable investment casting procedure for titanium alloys are (i) an energy-efficient method for melting titanium alloys under a controlled atmosphere that can handle high melt volumes, provide good homogeneity of alloys and enable a high superheat, and (ii) a non-reactive material that is conducive to making a viable mould. A material that corrodes during melting may still be usable as a mould material because of the shorter contact time and lower temperature during casting compared with melting (Freuh et al. 1996). Current melting practices of titanium have evolved to circumvent issues like temperature needed, reactivity of titanium, its sensitivity to interstitial elements, and the unavailability of a sufficiently inert crucible and mould.

4.25.2 INDUCTION MELTING

Induction melting is considered attractive, among other reasons, because the mixing it creates is good for producing homogeneous ingots. The crucible, if also used as susceptor, is at or lower than the melt temperature. Induction melting in an uncooled ceramic crucible provides accurate metal temperature control, melt thermal homogeneity and any desired molten metal superheating. A few of the titanium alloys are commercially melted by this method. But across the board, the use of this process is possible only when the crucible problem is solved. The problem remains one of the most actively investigated areas of titanium metallurgy.

Experimental results have often indicated that it is not thermodynamically possible to completely predict the inertness of a crucible material to attack by molten titanium. The general limits of theory in choosing the right crucible for melting titanium were described in Chapter 2. For checking on the possible reactions between potential crucible materials and molten titanium, numerous melting experiments are required.

4.25.3 SKULL MELTING

A good workable interim method for melting titanium without contamination is by the use of a water-cooled copper container to hold the melt. This arrangement is used in three major melting methods: vacuum arc melting, EBM, and ISM, or cold crucible induction melting (CCIM).

The real crucible is generally a water-cooled copper container. As the titanium is melted in this container, the melt immediately in contact with the water-cooled crucible wall freezes. From then on, the layer of titanium that is frozen acts in practice as the crucible inside wall in contact with the molten metal. The frozen layer is called the *skull*, and the melting style is known as *skull melting*. Skull melting can be implemented using a variety of heat input modes: by arc, induction or even electron beam, as noted earlier.

4.25.3.1 Induction Skull Melting or Cold Crucible Induction Melting (CCIM)

Cold crucible induction melting uses a segmented water-cooled copper crucible, and the melt is directly coupled to the high-frequency power supply. ISM of titanium scores well in alloy composition uniformity and is free from any crucible contamination. It is good enough for the production of small castings, because the technique is severely restricted due to lack of superheat control and the batch size limitation. Much of the applied energy to achieve optimal melt superheating is lost to the cooling water of the segmented crucible.

4.25.3.2 Vacuum Arc Melting (VAR)

Schuyler et al. (1976) summarized the status of titanium alloy melting and casting as regards various crucibles and moulds in the mid-1970s. Titanium casting producers used skull melting in a water-cooled copper crucible by the consumable and/or non-consumable electrode method, and for moulds, the choices were ceramic investment mould, using a non-reactive face coat, and rammed graphite moulds. In Russia (Glasunov 1970), consumable electrode melting into a graphite-lined, water-cooled copper jacket and casting in moulds fabricated from steel, graphite, alumina or sillimanite (Al_2SiO_5) lined with a graphite dispersoid was practised.

Most of the titanium produced today is arc melted in water-cooled copper crucibles under an inert atmosphere. Either non-consumable electrodes of graphite or tungsten or consumable electrodes of titanium are used. Consumable electrode arc melting in water-cooled moulds is a working alternative, but only a fraction of the metal is molten at any time, and the superheat margin is narrow. This condition is not conducive to alloy homogeneity, and multiple remelting is usually done to obtain homogeneous ingots. Even though arc melting methods must be considered as expedient, induction melting using an uncooled crucible remains the goal.

4.25.3.3 Electron Beam Melting (EBM)

EBM has been known since the 1950s and extensively used for refining refractory and reactive metals. More recently, this process has been used to produce alloys such as Ti6Al4V and very clean superalloys (Winckler and Bakish 1971) as well as titanium alloys (Otubo et al. 2003a, 2003b)

4.25.4 CASTINGS

Titanium alloys are not easy to machine. Casting is the way to go. At present, the consumable electrode vacuum arc method is used. Once the titanium electrode has been consumed, there is no more heat input to the system, and melt cooling is fast and intense. The metal must, therefore, be poured into the moulds very quickly before solidification in the crucible sets in. Under these conditions, a centrifugal casting arrangement is used to move this metal into the moulds (Bridges and Hauzeur 1991).

Although titanium is cast under vacuum and in a mould made of an appropriate non-reactive material that current technology can provide, it is not yet possible to avoid contamination of the casting surface by oxygen, carbonaceous material or first-coat oxides from the mould surface picked up by the molten metal at the time of pouring. The resulting impurity-laden layer, known as *alphacase*, is very hard and can range in thickness from zero upto 0.6 mm. The alpha case (α-case) is the subsurface layer that forms on a titanium casting in the area that is in contact with the mould. It is a thin, impurity-laden layer that develops due to mass transfer of elements such as oxygen, carbon, nitrogen, aluminium or silicon from the ceramic mould into the alloy as it solidifies (Saha and Jacob 1986). There is always a concentration gradient of elements across the thickness of the α-case, which has a break at the point where this layer ends. The α-case significantly decreases ductility, fracture toughness and fatigue life of the casting. There are various methods for removing it, including grinding, laser irradiation, water jetting, cathodic de-oxygenation and chemical milling in mixed acids. Yet, the better option would be to prevent or limit it rather than treat it. The best case is no case (layer). A mould material resulting in a thinner α-case is better than one that gives rise to a thicker α-case.

4.25.5 MOULDS

The methods for making moulds are: ceramic shells made by the lost wax process for investment casting to make complex shapes with a good surface finish, and precision sand and graphite mould-rammed sand, generally used for simpler shapes and with a surface finish slightly inferior to that obtained on investment castings. Standard investment casting moulds in general are not acceptable for casting titanium alloys. Zircon, alumina and silica are better avoided in the melt-facing surface due to their reactivity with molten titanium.

Shortlisting of materials for investment moulding can be done from crucible data. In any case, a mould is in contact with the reactive metal for only a short period, unlike a melting crucible, which contains a melt for a considerably longer duration. However, materials that may undergo dimension changes in firing is still acceptable for crucibles but not for moulds (Freuh et al. 1996, 1997; Kuang et al. 2000, 2001). Crucible investigations have identified several ceramics that possess low reactivities with the titanium alloys. These ceramics have been considered for mould applications also.

Ti-48Al was induction melted in a yttria-coated YSZ crucible at 1600°C and centrifugally poured into cylindrical moulds with face coats of zirconia, yttria and silica

– all under argon (Barbosa et al. 2006). Contamination occurs only at the casting surface, to a depth depending upon the mould material and solidification time. This is in addition to any α-case. The main contaminants for the moulds listed are oxygen and the element present in the oxide used as mould material: Si, Zr, Y. The best results were achieved in a mould made of zirconia with an inside coat of yttria. The α-case was only 25 µm, and there was no apparent metal–mould reaction. It is possible to cast titanium aluminides with very low contamination and good surface finish if yttria is used as face coat in the moulds. Plain zirconia moulds resulted in the formation of a ~15-µm thick Al_2O_3 layer on the sample surface (the result of a fairly complex sequence of reactions as described by Barbosa et al. 2006). The α-case was 75 µm thick. The silica mould resulted in the formation of a complex oxide layer, 20–50 µm thick, and the α-case was 250 µm in this situation.

Earlier, Schuyler et al. (1976) had observed that Ti-2.5Be, induction melted and cast in the $Y_2O_3 \cdot 15Ti$ crucible, contained oxygen levels (0.2–0.4%) that did not severely degrade cast alloy ductility. Yttria second-phase particles were observed in these castings. Castings made with the experimental Y_2O_3/K_2SiO_3 face coat mould and the Howmet Monograf moulds showed acceptably low surface contamination.

The possibility of significant preheating (1100°C) of ceramic moulds in the conventional vacuum investment process decreases the temperature differential between the mould and the molten alloy and enhances casting quality and reliability. This procedure is not yet feasible in a routine manner for commercial titanium alloys. The issue is materials. New materials are only just coming into view and are yet to be accessed and satisfactorily tested.

4.25.6 CRUCIBLES FOR TITANIUM METAL

Kroll himself started the search for a crucible to melt titanium when he investigated CaO as the crucible material in the year 1940. Investigations on ceramic crucibles suitable for titanium melting and reviews of various workable options started appearing regularly in publications beginning in the late 1940s. The first was an elegant paper by Brace (1948), who examined alumina, beryllia and thoria as possible candidates for a crucible. Among the three, thoria was judged to be the best. About a decade later, after considering many alternatives, Weber et al. (1957) suggested oxygen-deficient ZrO_2 as the optimal crucible material.

Beginning then, and during the seven decades that have since elapsed, the crucible materials that have been investigated at one time or other and for one reason or other have included refractory compounds of every description: nitrides, borides and carbides (Chapin and Friske 1954a, 1954b, 1955; Samsonov et al. 1960; Weimer 1997; Kartavykh et al. 2009, 2010; Sadrnezhaad and Raz 2005); sulphides (Eastman et al. 1951); oxides (Kuang et al. 2000; Sakamoto et al. 1992; Tetsui et al. 2012a, 2012b; Zhang et al. 2012); fluorides, oxyfluorides, graphite and refractory metals (Savitskii and Burkhanov 1970; Zhang et al. 2006; Frenzel et al. 2004; Nayan et al. 2007); as well as certain perovskites. However, the issue of crucible materials for induction melting of titanium and its alloys is still not fully solved, and no solution that has been put forward has evoked sustained confidence for significant commercial trials (Fashu et al. 2020).

4.25.6.1   The Oxides: Al_2O_3, BeO, ThO_2, CaO, MgO

When titanium was melted in contact with Al_2O_3, BeO and ThO_2 in vacuum, molten titanium reacted vigorously with Al_2O_3, less so with BeO, and slightly with ThO_2 (Brace 1948). Among these three oxides, only thorium oxide showed any promise as a crucible material for melting titanium or titanium-base alloys.

The most spectacular case was that of Al_2O_3. The interaction was so rapid that a visible increase of temperature occurred (something that could be observed in real time in the pyrometer used for sighting the charge and measuring its temperature) as soon as the titanium melted, and the titanium became so highly alloyed that the product bore no resemblance to either of the metals – titanium or aluminium. A titanium–aluminium intermetallic of sufficient stability could easily upset the stability order suggested by simple Ellingham plots. Later, Economos and Kingery (1953) noted that titanium melt could penetrate along the grain boundaries of Al_2O_3 at 1800°C, leading to the alternation of these oxides. The specimen melted in contact with BeO was also embrittled, indicating that some transfer of Be to titanium occurred. With thorium oxide, there appeared to be no significant interaction. Although it cannot be regarded as inert to molten titanium, it was potentially a suitable crucible material. Al_2O_3 and BeO would be useless as crucibles for melting titanium metal (Brace 1948).

Kroll (1940, 1956, 1959) stated that melting titanium on ThO_2 may be useful for commercial purposes if the presence of small traces of oxygen due to ThO_2 reduction is not objectionable. Titanium melted very quietly on a thoria base. When melted on an alumina, zirconia or yttria base, titanium readily reacted to form alloys. The mention of yttria along with alumina and zirconia is surprising, considering that many later investigations proved yttria to be largely inert to titanium. It needs to be taken into account that in 1940, suitable techniques for the separation of rare earths from one another were still in the early stages of development, and pure rare earth oxides, including pure yttria, were not readily available.

Pure ThO_2 crucibles fired at temperatures as high as 1860°C were also wetted by the titanium melt (displayed a concave meniscus; Weber et al. 1957). The Ti melt in the ThO_2 crucible had a dull, grey surface and a duplex microstructure resulting from solution of the crucible material in the melt and reprecipitation.

Schuyler et al. (1976) evaluated many crucible materials, and the results are listed in Table 4.4. They used a standardized method of evaluation. A 10-gram charge of the Ti–2.7Be alloy was melted in vacuum (0.1 to 0.01 Pa) in the crucible and held for 10 minutes at 1577°C (i.e. 150°C above the alloy melting point or at 150°C superheat).

Lyon et al. (1973) noted that hypostoichiometric yttria as a crucible resulted in less oxygen contamination in the melt compared with a fully stoichiometric oxide crucible. A similar suggestion was earlier made by Weber et al. (1957) regarding zirconia. To promote low oxygen stoichiometry in the crucibles, ThO_2 crucibles were coated on the inside surface with a carbon–acetone slurry, dried, and then re-fired at temperatures greater than 1615°C (Schuyler et al. 1976). When the crucible material was a substoichimetric ThO_2 (i.e. oxygen-deficient thoria of ThO_{2-x}

TABLE 4.4
Performance of Various Crucibles in the Vacuum Melting of Ti-2.7Be Alloys

Crucible	Increase in oxygen content, wt.%	Increase in other element content, wt.%
CaO·15Ti	1.06	Ca: 0.004
ThO₂	0.57	Th: 0.63
CaO (in inert gas)	0.59	Ca: 0.37
MgO (in inert gas)	3.41	Mg: 0.004
Y₂O₃	0.09	Y: 0.19
Y₂O₃.15Ti	0.12	Y: 0.19
SiC	Decrease of 0.02	Si: 7.51C: 5.85
CeS	0.03	Ce: 0.56S:0.19

Source: Schuyler et al. (1976).
Temperature: 1577°C (150°C more than the alloy solidus), holding time: 10 minutes, vacuum: 0.01 to 1 Pa.

or hypostoichiometric thoria) instead of fully stoichiometric ThO_2, the oxygen contamination of melts was 50% less.

Kroll (1940) noted that CaO is a satisfactory refractory for melting titanium but recommended that it be first fired near its melting point in a tungsten crucible in a good vacuum so as to crystallize it. Even this refined, crystallized CaO resists the action of molten titanium metal only at the metal's melting point. At temperatures slightly above this, a violent reaction occurs between titanium and calcia, resulting in the evolution of calcium vapour. Argon was used as cover gas for melting Ti alloys in CaO and MgO crucibles, because these crucibles show excessive metal boiling and spitting during melting in vacuum.

In regard to calcia crucibles and titanium melt, Schaffoner (2020) noted that calcium oxide, which decomposes to oxygen and calcium, shows fast dissolution with commercially pure Ti and Ti6Al4V due to the high vapour pressure of calcium. Calcia crucibles are good for titanium aluminide melts generally due to a much lower melting temperature of only 1460°C, calcium aluminate formation and possibly a lower titanium activity in the alloy (Morscheiser et al. 2008).

To assess CaO as a refractory material for titanium melts, its decomposition reaction has to be considered. At higher temperatures, calcium oxide decomposes to calcium vapour and oxygen, $CaO \rightarrow Ca(g) + 0.5O_2(g)$, a process noted by Kroll himself. A strong evaporation of calcium during the melting of titanium alloys was often reported, while the oxygen is picked up by the melt. As expected, the pick-up of calcium in titanium is very low, because most of it evaporates and does not remain there to be picked up. Calcia, therefore, lacks corrosion stability (Schaffoner 2020) with many low-alloyed titanium melts. Economos and Kingery (1953) observed that titanium melt could penetrate along the grain boundaries of MgO at 1800°C, leading to the alternation of these oxides. The situation as regards BaO is similar to that of CaO. There is also evaporation of BaO itself.

In their review, Weber et al. (1957) mentioned the results of titanium melting in some unusual crucible materials – titanium oxide, certain oxyfluorides and intermetallics. None of these could serve as crucible material for melting titanium.

4.25.6.2   Zirconia

The possibility of using a zirconia crucible for melting titanium was checked early on by Eastwood and Craighead (1950), who reported on many other crucible materials as well. These materials were refractory metals (W, Mo), carbon, carbides, oxides, nitrides, sulphides, borides and silicides. They considered that all materials other than zirconia were quite soluble in the titanium melt, and stabilized zirconia was the only refractory not wetted by the melt. The meniscus was convex. Oxygen was absorbed by the melt, but the Zr content of the ingot was generally low, hinting at the possibility that ZrO_2 was partly reduced and that lower oxides of zirconium might perform better as crucibles. However, the stabilized zirconia crucible failed as the ingot cooled. Economos and Kingery (1953) observed that titanium melt could penetrate along the grain boundaries of zirconia.

Zirconia stabilized with 15.5 mole% MgO and fired in an oxidizing atmosphere to maturity at 1860°C was used to melt titanium (Weber et al. 1957). A significant reaction occurred, and the melt stuck to the crucible. MgO-stabilized ZrO_2 crucibles blackened (indication of stoichiometry change of the oxide) during the melting operation, and the melt appeared to have diffused through the crucible wall.

Avoiding the alkaline earth oxides to stabilize zirconia, Weber et al. (1957) investigated ZrO_2 made oxygen deficient by reacting ZrO_2 with Ti metal powder. Titanium amounting to 15 atomic% was incorporated into ZrO_2 to make the crucible more resistant to attack by titanium melt. According to Ruh (1963), titanium could react with zirconia, and up to 10 at .% of Zr and O were taken into the solid solution of Ti, while the zirconia transformed into oxygen-deficient zirconia (Ruh and Garrett 1964). Zirconium entered the titanium lattice substitutionally, and oxygen went to interstitial positions. Lyon et al. (1973) pointed out that the free energy of formation of a Ti–10 at.% O solution at 1400°C was more negative than that of ZrO_2. The stoichiometry change in zirconia should not be surprising. Saha et al. (1990) also stated that ZrO_{2-x} and α-Ti(O) were formed when Ti reacted with ZrO_2.

When titanium was melted in 5 mol.% $CaO–ZrO_2$ or 3 mol.% $Y_2O_3–ZrO_2$ at 1750°C, the events were similar (Lin and Lin 1998, 1999b). Interfacial reactions between zirconia and titanium were preceded by the infiltration of liquid titanium through the open pores of zirconia. Zirconia was reduced to oxygen-deficient zirconia (ZrO_{2-x}), and the oxygen released was dissolved in titanium. The stabilizing oxide was expelled from zirconia. The Ca content increased in the solid solution of ZrO_{2-x}, leading to the precipitation of $CaZrO_3$. This may be contrasted with the fate of calcium when titanium was melted in pure calcia crucible. The yttria stabilizer was expelled as oxygen-deficient yttria. Saha et al. (1990) had made similar observations earlier.

4.25.6.3 Rare Earth Oxides

The possible use of rare earth oxides as crucible materials for melting titanium was investigated by Lyon et al. (1973) and Schuyler et al. (1976). The rare earth oxides were screened for their possible interaction with titanium in the solid state itself at 1000°C for 64 h under vacuum by setting up diffusion couples. Among the oxides, Y_2O_3, Er_2O_3, Ho_2O_3, Gd_2O_3 and Dy_2O_3 were the least reactive. The reactivity studies in molten condition were carried out in an arc melting furnace, melting the metal placed over the ceramic disc in the copper hearth. Induction melts were made at 1760°C under 3 mPa. Slight reaction in the form of some dissolution and precipitation of small dendrites close to the interface was shown by Pr_2O_3, Gd_2O_3 and Dy_2O_3. No dendrite precipitation occurred with yttrium oxide, the most stable of all.

Schuyler et al. (1976) fabricated crucibles of Dy_2O_3, HREMO (heavy rare earth mixed oxide), LREMO (light rare earth mixed oxide) and yttria for use as crucible and mould with titanium alloys. HREMO was the concentrate obtained from Xenotime, and LREMO materials were concentrates of Basnasite. The crucibles were reasonably dense (85–97% of theoretical density) and well bonded. The oxygen pick-up by titanium melt from these crucibles is given in Table 4.4. HREMO crucibles turned out to be second best only to Y_2O_3 and Y_2O_3·15Ti. Low reactivity was displayed by Dy_2O_3 also. Lowering the apparent oxygen stoichiometry was more effective in reducing oxygen contamination of melts made in HREMO (by nearly 50%) than in the case of Y_2O_3.

4.25.6.4 Yttria

Yttrium and oxygen ingested in titanium caused Y_2O_3 precipitation at the grain boundaries of the Ti matrix on cooling (Lyon et al. 1973). Like zirconia, on contact with titanium, Y_2O_3 can be reduced to Y_2O_{3-x}. The darkened colour of yttria revealed this. Melt experiments were performed using a sub-stoichiometric yttria crucible.

Yttrium oxide (Y_2O_3, yttria) is one of the most widely investigated refractory materials for titanium metallurgy, but its performance is still not considered completely satisfactory. Depending on the melting conditions, and possibly the crucible build, an oxygen contamination of 0.06–1.6 wt.% and a yttrium contamination of 0–1.93 wt.% occurs. There is also formation of yttria inclusions, attributed to the erosion of the crucibles by the melt (Schaffoner 2020).

Somewhat similar values were given earlier by Griesenauer et al. (1972). The oxygen and yttrium contamination in titanium melted in Y_2O_3 at 1770°C for 8 min under argon was 1.05 and 3.6 wt.%, respectively. Replacing the crucible with Y_2O_{3-x} at the same temperature under vacuum brought the contamination level O and Y down to 0.6 and 2.2 wt.%, respectively, even after prolonging the melting to 10 min. Y_2O_{3-x} is more resistant to attack by molten Ti. The 97% dense yttria crucibles experience thermal shock issues when the metal couples directly with the field in an induction furnace. The conditions become more manageable if a susceptor, like a molybdenum crucible, is used.

If titanium melt is to be contained in yttria, the best chance to minimize contamination is provided by the sub-stoichiometric yttria crucible. However, during cooling, dissolved yttrium scavenges oxygen dissolved in the melt and precipitates out as yttria particles. Incidentally, the presence of yttria particles results in relative softening of the titanium matrix.

Lyon et al. (1973) also observed that hypostoichiometric yttria as a crucible leads to decreased oxygen contamination in the melt. To promote low oxygen stoichiometry in the crucibles, Schuyler et al. (1976) used yttria crucibles re-fired (>1615°C) in argon by induction heating of a graphite susceptor. The crucibles were coated on the inside surface with a carbon–acetone slurry and dried. No carbon coating was given to Y_2O_3–15Ti crucibles.

Schuyler et al. (1976) summarized his observations as follows. (i)$Y_2O_3$is the best material for limiting contamination (0.16%oxygen) of low-melting titanium alloy melts. (ii) Lowering the apparent oxygen stoichiometry was not of significant benefit for Y_2O_3, whereas it was effective in reducing oxygen contamination of melts made in hypostoichometric HREMO and ThO_2 (by nearly 50%).

Three reports were issued by the Naval Research Laboratory, authored by Chapin and Friske (1954a, 1954b, 1955), covering oxides, carbon, graphite, carbides, borides and sulphides as possible crucible materials for melting titanium. The general conclusion was that all the materials investigated lacked the required inertness, as all reacted with Ti to varying degrees of severity.

4.25.6.5   Carbon, Graphite and Carbide Crucibles

Carbon and graphite are immediate natural choices for a crucible material for any melting in a non-oxidizing atmosphere. Chapin and Friske 1954b used crucibles of high-purity carbon with exceptionally smooth, dense surfaces and of commercially pure and spectroscopically pure graphite to melt titanium, specifically to check-whether purity and surface finish could make a difference. They did not. Crucibles made of the monocarbides of Ti, Zr, V, Nb, Ta and W were also investigated for their suitability to contain and melt titanium. Each of the carbides was wetted by molten titanium, resulting in general dissolution of the crucible and contamination of the melt, mainly with carbon. TiC was found to be the main carbide phase found in the solidified melt, formed from reduction of the crucible carbide by molten titanium at the liquid–solid interface. The carbides could not be considered suitable crucible materials. The results reported by Weber et al. (1957) and Lyon et al. (1973) in later years were similar to the conclusions of Chapin and Friske (1954).

Schuyler et al. (1976) evaluated SiC for possible use as crucible and mould for titanium alloys. The oxygen contamination of the alloy decreased, but there was a marked increase in Si and C contents. Slip casting a layer of Y_2O_3 on the inside of a SiC crucible was tried with the aim of combining the inertness of Y_2O_3 and the excellent thermal shock resistance of SiC. But, Y_2O_3 reacted with SiC during sintering, which converted the Y_2O_3 to a carbide. Further, SiC backup crucibles underwent structural deterioration.

4.25.6.6   Borides

Crucibles made of diborides of Ti, Zr and Cr were investigated by Chapin and Friske (1955) for melting titanium. In all cases, there was a general solution attack by molten titanium on the crucible, and the melt became severely contaminated with boron compounds. Solution of the reaction products raised the melting point of the melt, making pouring difficult. These borides are not suitable as crucible materials.

Titanium melts quietly in all the boride crucibles with no visible evidence of reaction between crucible and melt and no detectable vapour evolution. The affinity between titanium and the borides is so high that the reaction between the crucible and the metal is initiated in the solid state itself. All the borides were wetted by molten titanium. All the melts solidified in boride crucibles showed deep meniscuses and an acute angle of contact between crucible and melt. The melts became viscous soon after the charge melted and could not be completely emptied by pouring. The boride crucibles had good thermal shock resistance and did not crack either during the introduction of the somewhat cooler metal charge into a hot crucible or during melting or after pouring. In spite of their apparently excellent high-temperature properties, the borides lack the necessary chemical inertness toward molten titanium and are useless as crucible materials.

4.25.6.7 Sulphides

There are a limited number of sulphides with high melting points (above 1800°C) and good high-temperature stability. Two such compounds, CeS and ThS, have been considered for use as crucibles. The CeS crucibles were made by hydrostatic pressing followed by sintering at about 1950°C in vacuum. A CeS crucible was used for melting titanium.

Successful experimental fusions were reported (Eastman et al. 1951) for titanium in CeS crucibles. Other than the statement that a sound ingot was obtained without attack on the crucible (although the ingot adhered to the crucible), no details of metal quality were presented. Chapin and Friske (1955) melted titanium in a CeS crucible using induction heating and high vacuum. Melting occurred quietly. The molten metal poured without much difficulty provided it was not superheated excessively and was not held for extended periods. In either of these cases, it became viscous and difficult to pour. The resulting ingot, strongly yellow-coloured on the surface, adhered to the crucible and would not come off. All the metal melted in CeS crucibles was found to contain appreciable amounts of sulphur, ranging from 0.165 to 0.570 wt.%. Inspite of the ingestion of sulphur and cerium, the hardness of the metal was not markedly increased. This situation is similar to carbon ingestion into vanadium.

CeS is not sufficiently inert to prevent gradual consumption of the crucible by direct solution into the molten metal, but this material had the least deleterious influenceon the quality of metal melted in any of the materials evaluated in crucible form. Following Chapin and Friske (1955)'s report, Schuyler et al. 1976 melted a Ti–2.7Be alloy in a cerium sulphide crucible. The solidified alloy showed an increase in oxygen by 0.03 wt.%, cerium by 0.56 wt.% and sulphur by 0.19 wt.%.

4.25.7 CRUCIBLES FOR TITANIUM ALLOYS

Almost all ceramic crucibles, including Y_2O_3, are attacked by molten titanium to a certain extent. In the melting of TiAl alloys, Y_2O_3, $BaZrO_3$, Ca-doped $BaZrO_3$ and

AlN are the most resistant to attack, and a final <0.1 wt.%oxygen content in the ingot can be achieved.

4.25.8 TiAl Alloys

The technologies for manufacturing billets of TiAl-based intermetallic compounds are melting and casting in a ceramic mould and directional solidification in a ceramic crucible under inert gas or vacuum. Oxygen contamination is, in general, hard to control but needs to be controlled. Oxygen gains entry into a melt from cover gas and especially from insufficiently inert crucible and mould. The elements that are part of the oxide ceramics, i.e. Y, Zr, Ca, Mg and Si, also diffuse from the crucible or mould walls into the ingot. Fashu et al. (2020) have summarized, in Table 4.5, the current status as regards the use of various crucibles for the melting of titanium–aluminiumalloys.

4.25.8.1 Zirconia (Stabilized) Crucible

TiAl alloys were induction melted (1500°C, 2 min) in argon using crucibles of zirconia stabilized with yttria, calcia or magnesia (Barbosa and Ribeiro 2000). All the samples picked up (~0.27 to 0.93 wt.%) zirconium in addition to oxygen. Of the three types, calcia-stabilized zirconia contaminated the melt least. The α-case was 300–600 μm thick. The contamination was less (α-case 200 μm) when pure calcia crucibles were used.

TiAl melted and cooled in graphite crucibles was contaminated with 3–7 atom % carbon. The larger value corresponds to regions near the surface. However, the alloy melted in yttria-stabilized zirconia and cast in graphite mould showed no sign of carbon contamination.

Lin and Lin (1999a) vacuum arc melted Ti-6Al-4V alloy and cast it centrifugally in a porous ZrO_2 mould. Interfacial reactions between Ti-6Al-4V and ZrO_2 mould were preceded by the penetration of liquid titanium due to capillary and centrifugal forces through the interconnected pores near the mould surface. Ti leached oxygen from ZrO_2 to form α-Ti(O) solid solution. Zirconium also dissolved in α-Ti(O). The SiO_2 in the binder in the ZrO_2 mould readily reacted with titanium to form Ti_5Si_3. The stabilizer CaO incorporated into zirconia is released, dissolves in Ti, and reacts with ZrO_2 and Ti(O) to form $Ca_3Ti_2O_7$ and $CaAl_4O_7$ in the reaction zone.

Hypostoichiometric zirconia contaminates Ti melt less, not only in crucibles during melting but also in moulds during casting. Nakamura et al. (1988) incorporated zirconium powder (3–5 wt.%) in the zirconia face coat to serveas a barrier to the diffusion of oxygen from the mould into titanium. The zirconia face coat, made hypostoichiometric by vacuum firing, also resulted in a thinner α-case (Calvert 1981).

Gao et al. (2011) investigated the effect of porosity (16–42%) of Y_2O_3 on the contamination levels of Ti-54Al (at .%) melted (1600°C, 100 min) and solidified in the crucible. Depth of melt penetration, oxygen pick-up, amount of yttria particles and also yttrium content increased with increase in porosity and holding time. The oxygen content of the original melt stock was 400 ppm. The final oxygen content in the alloys made by Gao et al. (2011) using a yttria crucible of low (16%) porosity was

TABLE 4.5
Contamination of TiAl Alloys during Vacuum Induction Melting in Different Crucibles

Crucible material	Holding time, min	Interaction layer, μm	Oxygen content, wt.%	References
SiO_2	0.5–5	Not reported	1.7, Si:1.7	Tetsui et al. 2012b
Mullite $(3Al_2O_3 \cdot 2SiO_2)$	15–60	190–840	–	Kuang et al. 2001
MgO	18–60	1296–3570	0.95–1.09	Kuang et al. 2000
Al_2O_3	0.5–60	30–247	0.16–2.5	Eatesami et al. 2009, Kuang et al. 2000, Kuang et al. 2001, Liu et al. 2005, Tetsui et al. 2010, Tetsui et al. 2012b, Zhang et al. 2013a, Zhang et al. 2012
ZrO_2 (incl. stabilized)	0.5–60	120–300	0.55–1.29	Barbosa and Ribeiro 2000, Kuang et al. 2001, Tetsui et al. 2010, Tetsui et al. 2012b, Zhang et al. 2013a
$Y_2O_3 \cdot 15ZrO_2$	3	Not reported	0.38	Tetsui et al. 2012b
CaO	0.5–60	5	0.10–0.71	Barbosa and Ribeiro 2000, Kuang et al. 2000, Liu et al. 2005, Sakamoto et al. 1992, Tetsui et al. 2010, Zhang et al. 2013a
Y_2O_3	0.5–60	8–100	0.06–0.33	Cui et al. 2012, Hockaday and Bisaka 2010, Kuang et al. 2001, Liu et al. 2005, Cui et al. 2010, Tetsui et al. 2010, Tetsui 2012b, Zhang et al. 2013a
Y_2O_3-coated MgO	18–60	10–19	0.016–0.21	Kuang et al. 2000
Y_2O_3-coated Al_2O_3	10–34	Not reported	0.04–0.13	Schuster and Palm 2006, Van Humbeeck 1999
Y_2O_3-coated ZrO_2 (incl. stabilized)	60–120	15–65	0.21–0.31	Barbosa et al. 2007, Barbosa et al. 2003, Barbosa et al. 2005, Barbosa et al. 2006
Y_2O_3-coated $ZrO_2 \cdot SiO_2$	1	Not reported	0.19–0.22	Gomes et al. 2013
$CaZrO_3$	5–20	300–350	0.13–0.73	Schafföner et al. 2015a, 2015b, Li et al. 2010, Klotz et al. 2019
$CaZrO_3$–Al_2O_3 composite	30	150	–	Lu et al. 2017

(Continued)

TABLE 4.5 (CONTINUED)

Contamination of TiAl Alloys during Vacuum Induction Melting in Different Crucibles

Crucible material	Holding time, min	Interaction layer, μm	Oxygen content, wt.%	References
BaZrO$_3$	5	870–2000	0.065–0.13	Chen et al. 2018b, Zhang et al. 2013b
Ca-doped BaZrO$_3$	2	5	0.09	Chen et al. 2018a
AlN	5–25	1.1–6.4	0.025–0.113	Chapin and Friske 1954b
BN	5–25	250–>500	–	Chapin and Friske 1954b
Graphite	0.3–2	Not reported	C:4	Barbosa and Ribeiro 2000, Cegen et al. 2016
Water-cooled Cu	3	0	0.09	Tetsui et al. 2012b

Source: Fashu et al. 2020.

not more than 800 ppm even after a holding time of 100 minutes. If this inertness is repeated in larger crucibles and bigger batches, and if the crucible survives the thermal shock resistance test, the crucible problem may be seeing the light at the end of the tunnel for TiAl alloy melting.

The impurities to look for in titanium–aluminium alloys are oxygen and carbon, and according to industrial standards, the oxygen content should not exceed 0.1 wt.% (Tetsui et al. 2012b). A summary of the performances of different crucible materials used to melt TiAl alloys (1650°C, 5 min, Ar) is shown in Figure 4.17A and 4.17B.

Among the oxide crucibles tested for melting titanium alloys, yttria and calcia were best in terms of chemical inertness, but they are both poor in shock resistance (Barbosa et al. 2007; Cui et al. 2010; Barbosa and Ribeiro 2005). Y$_2$O$_3$ meets the requirement of <0.1 wt% oxygen in ingots (Tetsui et al. 2012a, 2012b).

4.25.8.2 Yttria

The details of melting and consequent oxygen enrichment in the melt (+) are given in Table 4.6.

Every sample had dispersed yttria inclusions (20 μm) in the metal matrix. The average measured oxygen content of the melt charge was 0.08 wt.%. SEM X-ray dot map micrographs reveal that there was no penetration of the melt into the Y$_2$O$_3$ crucible, even though the latter contained crevices and cracks. The oxygen enrichment of the cast samples (0.06–0.10 wt.%) obtained by Cui et al. (2010) are consistent with those reported by Kuang et al. (2000) for TiAl melted in yttria-coated crucibles. The presence of small yttrium oxide particles dispersed in the metal matrix could be due to erosion promoted by inductive stirring action in the melt during melting,

Lapin et al. (2011) tested a Y$_2$O$_3$ crucible with Ti46Al8Nb (at .%) alloy (liquidus 1570°C) for directional solidification (DS) in a Bridgman-type apparatus under 10 kPa argon. The overall interaction was found to be diffusion-controlled dissolution

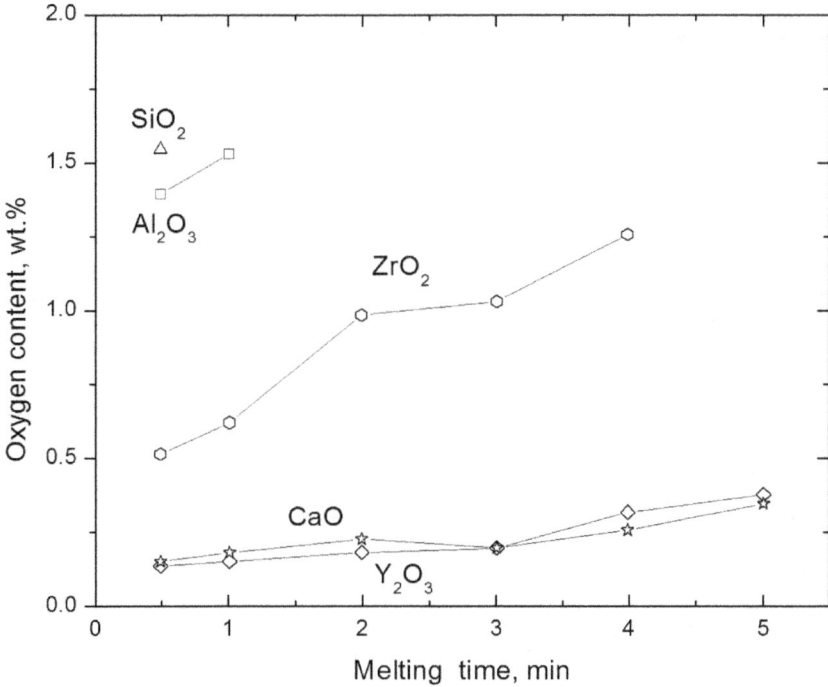

FIGURE 4.17A Melting time and O pick-up by TiAl alloys from different crucibles I (From Tetsui et al. 2012b).

of the Y_2O_3 crucible in the melt, resulting in an increase of oxygen and yttrium content in $\gamma(TiAl)+\alpha2(Ti3Al)$ matrix and precipitation of Y_2O_3 particles in the inter-dendritic region. The change in oxygen content with temperature and duration of contact in the DS samples (Lapin et al. 2011) at other temperatures also is given in Figure 4.18. On the whole, the ingestion of impurities appears controllable and the Y_2O_3 crucible useful for DS with this alloy. The crucibles used by Lapin et al. (2011) were dense (92.8% of TD or theoretical density) with very small (0.7) volume fraction of open porosity. An oxygen content of about 1200 wt.ppm in cast TiAl-based alloys is within the technically acceptable limit.

Barbosa et al. (2005) melted Ti-48Al alloy at 1600°C and cooled it inside CaO-, MgO- and Y_2O_3-stabilized ZrO_2 crucibles with an inner 200-μm thick layer of Y_2O_3. Every sample was found to have a small amount of yttrium oxide inclusions, possibly eroded from the coating. In the sample melted in the yttria-stablized zirconia crucible that was not coated with yttria, the inclusions were of alumina. Every sample was contaminated with a small amount of yttrium (average value between 0.14 and 0.16 at.%). No zirconium was detected in solution. The Y_2O_3 layer effectively isolates the crucible base material from the melt. The crucibles and coating were intact after melting. Significantly, the solidified cylindrical ingot of the alloy has a concave top surface when prepared in the uncoated crucible as compared with the distinct convex top surface of the alloy melted and cast in the coated crucible – a pointer to the

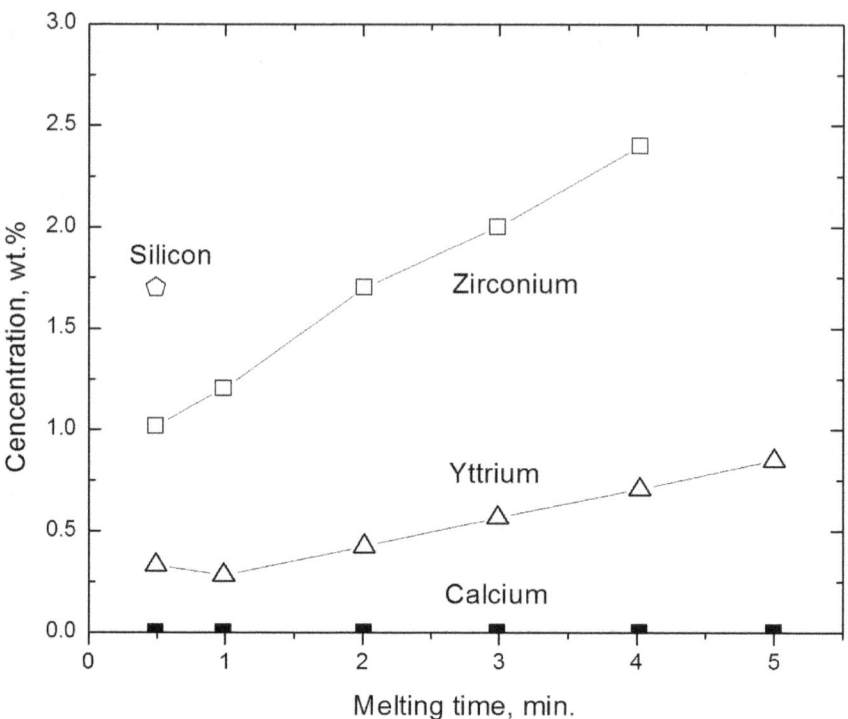

FIGURE 4.17B   Melting time and pick-up of other elements by TiAl alloys from different crucibles II (From Tetsui et al. 2012b).

significant difference in the wetting behaviour. Earlier, Spedding et al. (1970) had given a sketch representing the cross section of a typical reduction crucible in calciothermic reduction of rare earth trifluorides. Figure 4.15 highlighted the wetting behaviour of the reaction products and the shape of solidified ingot.

In a later investigation, Barbosa et al. (2007) applied the Y_2O_3 layer on a zirconia crucible by painting followed by a drying operation, and no sintering was done. The porosity was probably high and adhesion to the zirconia crucible lower than optimum. In spite of this, the absence of Zr in the cast alloy showed that the Y_2O_3 layer was an effective barrier between the crucible base material (ZrO_2) and the molten metal due to its limited wettability by TiAl (Kuang et al. (2001).

4.25.8.3   Calcia

Sakamoto et al. (1992) prepared Ti-33–38 wt.% Al alloys by VIM (1550–1620°C, 20 min) in a calcia crucible of cp-titanium sheets and cp-aluminium blocks. The charge (~7 kg) was coupled directly in the 125 mm dia crucible, and the melting was done in argon. The behaviour of impurity elements in the molten titanium aluminide is shown in Figure 4.19. Impurity content (Fe, Si, N and H) did not vary except for oxygen and calcium. The oxygen content of the alloy charge was 0.034 wt.%, and the

TABLE 4.6
Contamination from Crucible in TiAl Melting in Argon Atmosphere

Melt- crucible combination	Temperature, °C	Time, min	Contamination	Reference
Ti-47 at % Al-yttria	1600.	20	Oxygen: +0.06 to 0.07 wt.%	Cui et al. 2010
	1700	20	Oxygen: +0.08 to 0.10 wt.%Yttrium: +0.15 to 0.19 at .%	
Ti-46Al-8Nb (at.%) - yttria	1650	210	Oxygen: +0.4 at.%	Lapin et al. 2011
Ti-48 at% Al-ZrO$_2$ stabilized with CaO, MgO or yttria with 200 μm yttria coating	1600		Oxygen:Zirconium: +zeroYttrium: +0.14 to0.16 at .%	Barboza et al. 2005
Ti-33–38 wt.% Al-calcia (0.034 wt.% O, <0.01 wt.% Ca in the alloy)	1550 to 1620	20	Oxygen: 0.10 to 0.13 wt.%Calcium: −(decrease)	Sakamoto et al 1992

+ represents ingestion, − represents decrease.

calcium content was <0.001 wt.%. Oxygen content increased to 0.1–0.13 wt % in 20 min, and calcium content decreased.

4.25.8.4 Graphite

Barbosa and Ribeiro (2000) do not suggest induction melting of TiAl-based alloys in graphite crucibles because of contamination of products by carbon and carbides. However, Cegan et al. (2014) noted that a technically acceptable content of carbon-can be achieved in TiAl-based alloys by using better-quality graphite crucibles.

On induction melting of TiAl at 1500°C for 2 min, Barbosa and Ribeiro (2000) found the alloy to be contaminated with 3–7 atom % carbon. However, the alloy melted in yttria-stabilized zirconia and cast in a graphite mould showed no sign of carbon contamination. Cegan et al. (2015) prepared three different Ti-47Al-8 Nb/Ta alloys by induction melting (1700°C, 20s, Ar) of the components(Ti, Al, Y, Nb-60Al and Ta-80Al) in graphite crucibles and centrifugal casting into graphite moulds. The compositions are all in at.%. The carbon content measured in the as-cast alloys is relatively low when compared with that (3–7 at .%) reported by Barbosa and Ribeiro (2000). The lowest carbon content of 460 wt.ppm is found in the 8Ta- alloy, which corresponds to about 0.2 at.%. Cegan et al. (2015) did the melting at 1700°C for 20 seconds compared with 1500°C for 2 min. by Barbosa and Ribeiro (2000). A lower

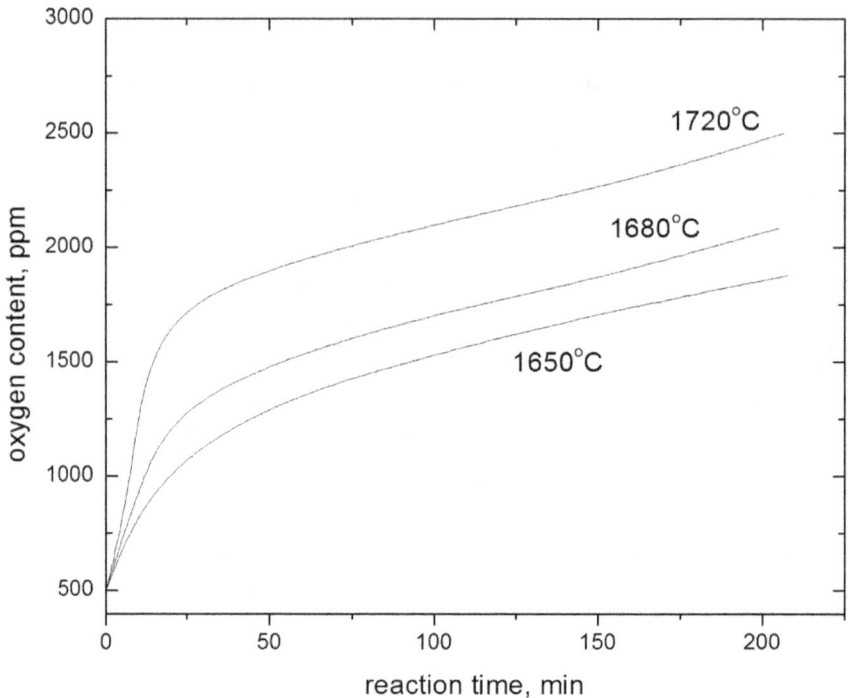

FIGURE 4.18 Oxygen pick-up during directional solidification of TiAl in yttria (From Lapin et al. 2011).

carbon ingestion in spite of the much higher temperature used may be traced to the quality of the graphite crucibles.

Melting in graphite crucibles and casting into graphite moulds can be considered as a viable method of preparation of TiAl-based alloys containing tantalum and niobium (Cegan et al. 2015).

In a later investigation, Lapin and Klimová (2019) prepared Ti-38 to 45Al alloys with Nb and Mo additives by VIM at 1700°C under 10^3 Pa argon for about 3 min in graphite crucibles followed by casting into graphite moulds. VIM in high-density graphite crucibles resulted in a reproducible increase of carbon content by (0.9 ± 0.1) at .%, which can be accommodated in the process by a suitable choice of starting alloy composition. Nevertheless, Fashu et al. (2020) have remarked that TiAl alloys are yet to be successfully melted in graphite crucibles keeping contamination levels sufficiently low as required by industrial standards.

4.25.8.5 Nitrides: AlN and BN

Both high-purity AlN and BN have been evaluated as crucibles for melting TiAl (Ti-46Al-8Nb) alloys (Kartavykh et al. 2009, 2010). The AlN crucible with the sample was heated to 1670°C (i.e. 100°C higher than the liquidus temperature), held for up to 25 min and rapidly cooled.

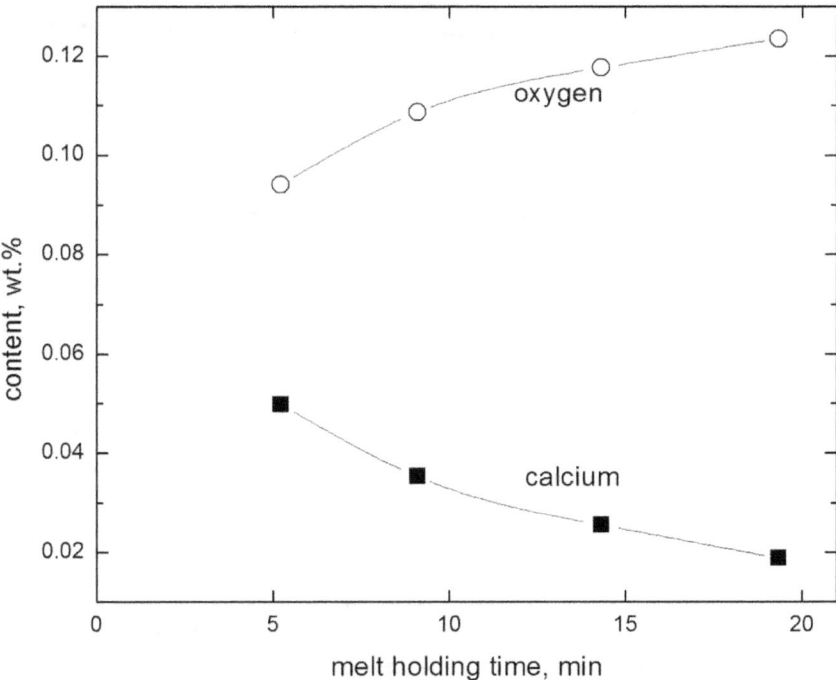

FIGURE 4.19 Impurity content in TiAl after melting in calcia crucible (From Sakamoto et al. 1992).

The free surfaces of all solidified melt samples were clean, without temper colours or signs of oxidation. The solidified casting easily came off from the crucible. On prolonged (25 min) superheating, micro-precipitates of nitride phases $((Ti,Al)xNy, NbN)$ formed in the $\alpha 2$ matrix phase $(Ti3Al)$ in the near-contact zone at a distance of no more than 300 μm from the ingot–crucible interface. However, Kartavykh et al. noted that this layer is removable during the finishing of as-cast parts.

The alloy ingots made in these experiments had no oxygen-containing (oxide) phases. Their maximum oxygen was 1100 wt.ppm, which is 1.5–2 times lower than that obtained in similar experiments performed on Y_2O_3 crucibles (Kartavykh and Cherdyntsev 2008). The oxygen ingestion on heating the titanium–aluminiumalloy in various crucibles is summarized in Figure 4.20.

Kartavykh et al. (2010) also used small pyrolytic boron nitride (pBN) boats to hold and melt Ti-46Al-8Nb (at .%) alloy. The facility and procedure were the same as those used for AlN. Boron nitride is wetted by the alloy melt. The melt had reacted with the boron nitride, forming $Ti2AlN$ and $(Ti,Nb)B$ in the form of needle-like micro-precipitates at the surface between the alloy and the crucible. The reaction layer comprising these particles propagates quickly into the alloy bulk, spreading to the whole sample in about 20 min of holding at temperature. pBN is unsuitable as a material for crucibles/moulds for titanium-based melts.

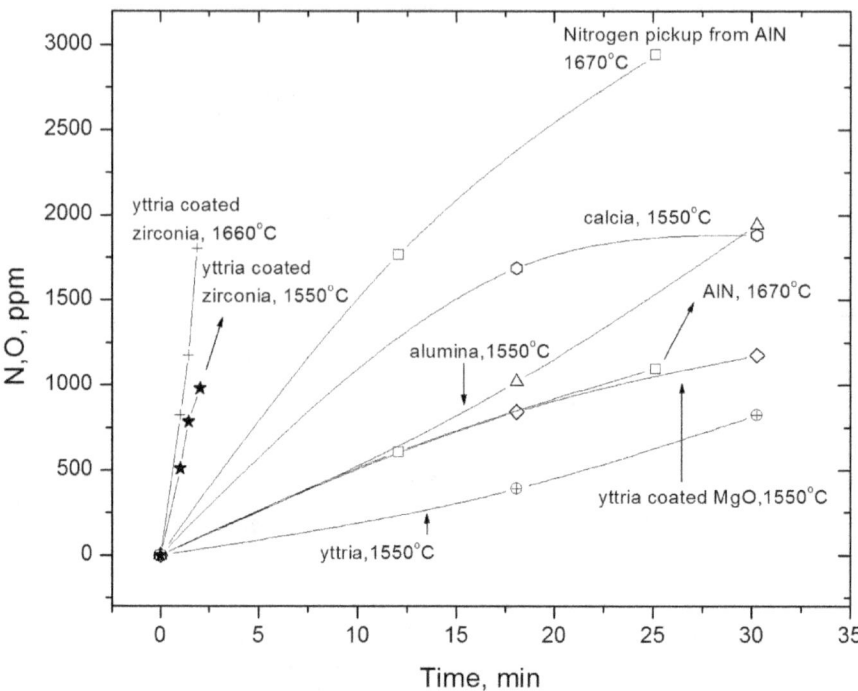

FIGURE 4.20 Pick-up of interstitials by TiAl after melting in AlN and other crucibles (From Kartavykh et al. 2009).

4.25.8.6 The Zirconates

It is seen that crucibles of oxides, carbides, borides and sulphides have been tested for melting titanium alloys. CaO and Y_2O_3 were best in chemical inertness but poor in shock resistance. CaO is also hygroscopic. A chemical inertness equal to or exceeding that of yttria and calcia, and a much better resistance to thermal shock, is possessed by the zirconates – $CaZrO_3$ and $BaZrO_3$.

4.25.8.6.1 Calcium Zirconate

When calcia-stabilized ZrO_2 was heated in contact with titanium in solid state and the interface was examined, $CaZrO_3$ had formed at the interface. Chang and Lin (2010) surmised that $CaZrO_3$ is a stable phase in contact with titanium alloys and a potential refractory material for titanium metallurgy. $CaZrO_3$ was evaluated in recent years in contact with commercially pure titanium (cp-Ti), Ti6Al4V, titanium aluminide (TiAl) and TiFe. In contact with Ti6Al4V, $CaZrO_3$ showed remarkable corrosion resistance when used as investment casting moulds. Compared with moulds based on Y_2O_3, the zirconate caused a much lower hardness increase and excellent demoulding (Freitag et al. 2017; Klotz et al. 2019). $CaZrO_3$ moulds were used by Kim et al. (2001) to cast commercial-purity titanium metal, Ti6Al4V and γ-TiAl melts. $CaZrO_3$ was comparable to CaO in stability.

Li et al. (2010) used calcium zirconate (CaZrO$_3$), perhaps for the first time, as a crucible for melting Ti6Al4V and also NiTi. Only slight corrosion was reported. SEM and X-ray examination of the solidified melt showed that both alloys formed an interfacial reaction layer with the ceramic, but there was no transfer of alloy or crucible components across the interface to the other side, indicating that CaZrO$_3$ is a promising refractory for melting titanium alloys.

Li et al. (2007, 2013) obtained coarse-grain CaZrO$_3$ by slip casting of CaCO$_3$ and ZrO$_2$, firing at 1500°C for 3 h and crushing. The product had 12.4% open porosity. Schaffoner et al. (2013) tried both solid-state synthesis and electric arc melting methods to prepare coarse-grained calcium zirconate. The fused calcium zirconate had considerably lower porosity than solid-state synthesized material. After crushing and milling of fused CaZrO$_3$, preparation of large crucibles by cold isostatic pressing resulted in crucibles with 15.9–16.5% porosity.

Schaffoner et al. (2015a, 2015b) tested the corrosion of fused CaZrO$_3$ crucibles in contact with γ-TiAl melts during VIM. The crucibles were prepared by cold isostatic pressing, dried and then fired at 1650°C for 6 h. The batch size was approximately 1000 g, and the molten mass was held for 20 min before casting in massive steel moulds. There was a uniform reduction in diameter of the crucible due to visible corrosion. The melts were contaminated with both zirconium (2.71–9.96 at.%) and oxygen (0.45–0.8 at.%). There was no stoichiometric correlation between the amounts of zirconium and oxygen taken up by the melt. Interestingly, calcium zirconate ingests more zirconium into the melt than pure zirconia (2.1–3.1 at .%) crucibles (Friedrich et al. 2008). The oxygen contamination was greater than what was observed after melting TiAl in CaO crucibles (<0.35 m.%). under similar conditions. The crucible was infiltrated by the melt. Beginning from the melt side, a reaction zone 2 mm thick was also observed. Fused CaZrO$_3$ did not prove to be a satisfactory refractory to melt pure γ-TiAl.

In contact with cp-Ti, fused CaZrO$_3$ reacted strongly, especially by dissolution (Schafföner et al. 2015a, 2015b). As mentioned, electrofused raw material contains a certain amount of c-ZrO$_2$ impurities resulting from evaporation of CaO, and this is responsible for the marked reactivity of fused calcium zirconate wares. Schaffoner et al. (2015a) tested crucibles with four different compositions – one plain and the other three CaO adjusted – in contact with Ti6Al4Vmelts. Melting experiments (1100 g batch, 1670°C, 20 min) were done with Ti6Al4Vin a vacuum induction furnace under backfilled argon (800 mbar) to suppress the evaporation of volatile elements such as calcium. The melt was cast in steel moulds.

Dissolution of the zirconia phase by the alloy melt occurred. The calcium zirconate phase did not dissolve. The zirconium and oxygen contamination of the melts, which was highest for the plain fused calcium zirconate crucibles that contained a substantial amount of c-ZrO$_2$, decreased progressively as the c-ZrO$_2$ proportion of the calcium zirconate crucibles was decreased by increasing the amount of calcium hydroxide additive in crucible making. The calcium hydroxide addition cannot be increased without limit, because after a certain level, the crucibles started disintegrating. When Ti6Al4Vwas melted in yttria crucibles under similar conditions (Friedrich et al. 2008), melt contamination by oxygen was 0.74 wt.% and by yttrium was 0.26

wt.%. Lower levels of oxygen contamination (0.43 wt.%) were reached in the composition-adjusted fused calcium zirconate crucibles (0.50 wt.% Zr).

A substantial infiltration by the melt occurred in fused calcium zirconate crucible with no Ca(OH)$_2$ added. The infiltration by the titanium alloy melt decreased to a hardly detectable level with calcium hydroxide additions. Possibly, the infiltration of the crucibles occurred by a preferred dissolution of the zirconia phase accompanied by a subsequent infiltration along the dissolved zirconia phase. It is known that zirconia dissolves into titanium even in the solid state (Ruh 1963; Lin et al. 1999b). The zirconate phase remained unaffected.

4.25.8.6.2   Barium Zirconate

Barium zirconate was evaluated as a crucible material for melting titanium, TiAl, TiNi, Ti2Ni and alsoTi6Al4V and cp-Ti. Melts richer in titanium significantly increase the corrosion of BaZrO$_3$ (Chen et al. 2019). The corrosion was suppressed by additions of CaO and Y$_2$O$_3$ to BaZrO$_3$. But, Y$_2$O$_3$ doping led to poor sintering, which could be offset by adding TiO$_2$. Compared with CaZrO$_3$, crucibles made of BaZrO$_3$ (Zhang et al. 2013b) did not result in an interaction layer, and the oxygen contamination was lower than in Y$_2$O$_3$ crucibles.

Chen et al. (2019) carried out pilot-scale melting of TiAl-based alloys using fused BaZrO$_3$ crucibles. Crucibles (inner diameter of 190 mm and length of 290 mm) were prepared by cold isostatic pressing using BaZrO$_3$ refractory of less than 3 mm average particle size. They were then fired at 1750°C for 6 h. About 25 kg of the components (titanium sponge, aluminium ingot Al-60 wt.% Nb master alloy to yield Ti46Al8Nb [at .%]) were charged in a BaZrO$_3$ crucible and melted (1550°C, 30 s) in a vacuum induction furnace under argon. The alloy melt was cast in a graphite mould.

The objective was to get a homogeneous Ti46Al-8Nb alloy with less than 0.1 wt.% oxygen (Tetsui et al. 2012a, 2012b). The average oxygen concentration of ingots was about 0.078 wt.% (i.e. a pick-up of approximately 0.05 wt.% oxygen), and the variation of oxygen content with sampling position in the ingot was marginal, indicating good homogeneity inspite of a very short holding time (30s) during induction melting in 25 kg batches using190 mm diameter crucibles. In experiments carried out on a similar scale using CaO and Y$_2$O$_3$ crucibles, the oxygen content in the alloys was 0.24 and 0.32 wt.%, respectively (Friedrich et al. 2008). BaZrO$_3$ appears to be a less contaminating refractory for induction melting of TiAl-based alloys (Chen et al. 2019).

The Ca-doped BaZrO$_3$ crucible was prepared starting from BaCO$_3$, ZrO$_2$ and CaO by cold isostatic pressing followed by solid-state sintering at 1750°C. The refractory consisted of Ba$_{1-x}$Ca$_x$ZrO$_3$ and CaO phases (Chen et al. 2018a, 2018b). Figure 4.21 shows the oxygen pick-up by the TiAl alloy in the Ca-doped BaZrO$_3$ and BaZrO$_3$ crucibles, along with the literature data for samples prepared in other oxide crucibles.

The objective stated by Chen et al. (2018c)was to produce TiAl ingot with a composition as close as possible to Ti-46 at.% Al. High-purity raw aluminium pellets (>99.99%) and sponge titanium (>99.9%) charge were heated in a vacuum induction furnace at 1600°C for 2 min under argon. The melt was either poured into a graphite

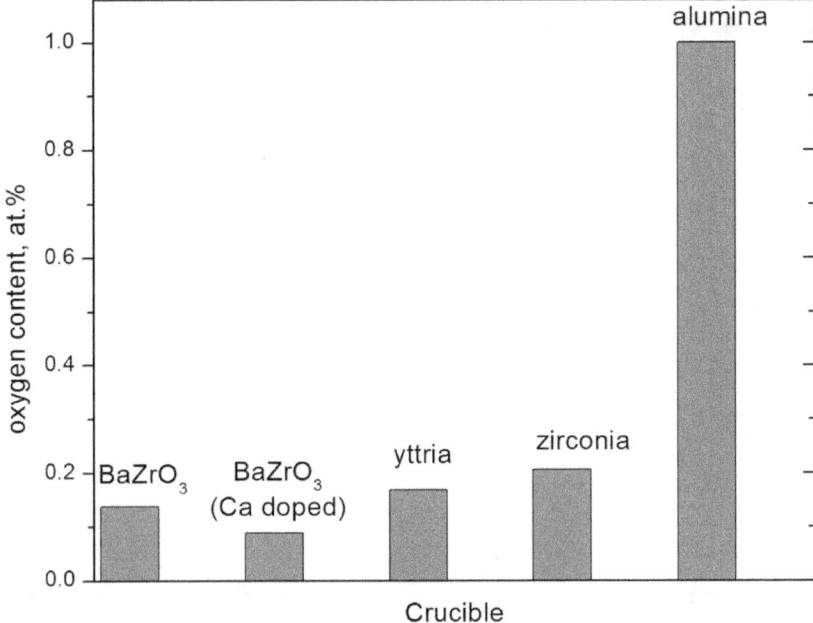

FIGURE 4.21   Oxygen pick-up by TiAl from calcium-doped barium zirconate (From Chen et al. 2018c).

mould or solidified in the the Ca-doped BaZrO₃ crucible. The oxygen concentration of the resulting TiAl ingot was 0.09 wt.%, lower than the maximum permitted oxygen concentration (Sakamoto et al. 1992). In comparison, the oxygen content of alloy melted in a plain BaZrO₃ crucible was about 0.13 wt.%.

The oxygen concentration of the TiAl alloy melted in the Ca-doped BaZrO₃ crucible (0.09 wt.%) was noticeably lower than for the other conventional crucibles such as Y_2O_3, ZrO_2 and Al_2O_3 as well as plain BaZrO₃, although the thermochemical stability of Y_2O_3 and CaO refractories was higher than that of the Ca-doped BaZrO₃ refractory (Chen et al. 2017). A possible explanation could be a difference in the wetting behaviour of these refractories. Ca-doped BaZrO₃ is a very promising candidate crucible for induction melting of the highly reactive TiAl alloys (Chen et al. 2018c).

4.25.8.7   CCIM

Cold crucible induction melting (CCIM) in a water-cooled segmented copper crucible was one of the two methods used by Sakamoto et al. (1992) to prepare Ti-33–38 wt.% Al alloys by melting together cp-titanium sheets and cp-aluminium blocks. Melting was carried out in a crucible that could hold about 30 kg of titanium aluminide. The temperature was 1550–1620°C and holding time 25 min – all under 2.7 × 10⁴ Pa backfilled argon. The oxygen content of the alloy charge was 0.034 wt.%, and the calcium content was <0.001 wt.%, which did not change in the titanium aluminide obtained. As seen in Figure 4.22, this was a melting truly free from crucible contamination.

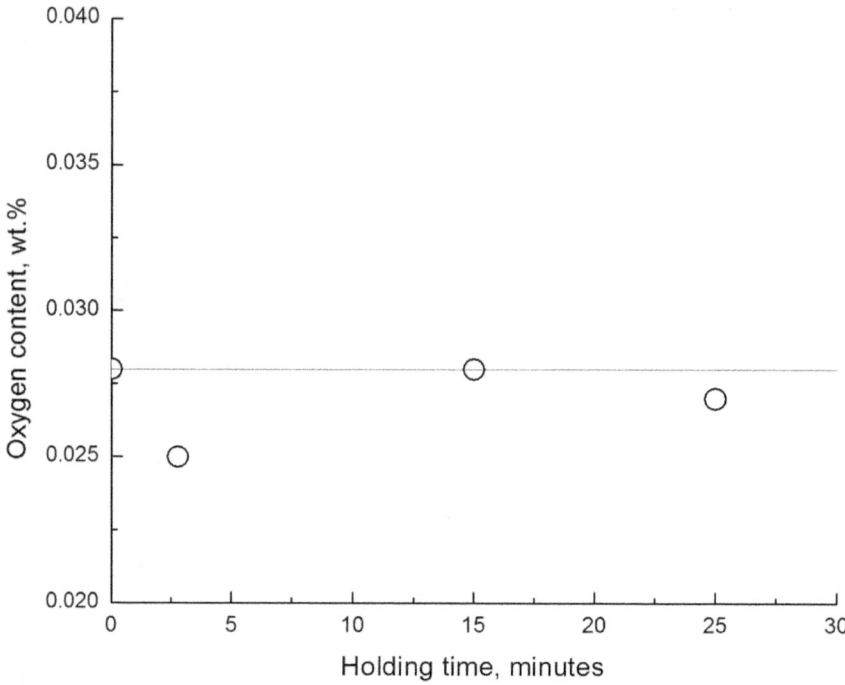

FIGURE 4.22 Change in oxygen content of TiAl by cold crucible induction melting. (From Sakamoto et al. 1992).

4.25.8.8 EBM

Apart from CCIM, Sakamoto et al. (1992) prepared Ti-33–38 wt.% Al alloys by melting together cp-titanium and cp-aluminium in an EBM furnace. The oxygen content of the charge decreased from 0.032 and 0.043 wt.% to <0.01 wt.%. Aluminium content also simultaneously decreased by 10 wt.%. The aluminium content is eventually adjusted in EBM by adding Al through the feeder.

4.25.9 NiTi(-X)-Based Shape Memory Alloys (SMA)

The conventional process to produce NiTi alloy is by VIM. The limitations are contamination of the bath by carbon from graphite crucibles or graphite moulds or by oxygen from the use of MgO and Al_2O_3 crucibles (Jackson et al. 1972). Ni50Ti melts at~1310°C. Alloy melting is generally performed between 1450 and 1500°C. These temperatures are lower than those required for melting TiAl alloys. However, the need to control composition within a very narrow margin near equiatomic composition does keep the pressure on the melter (Fashu et al. 2020). Besides, impurity ingestion can easily make the alloy useless for its intended applications. The shape memory effect and the related functional and strength properties of NiTi alloys are strongly dependent on both composition and purity of the alloys, When carbon

TABLE 4.7

Interstitial Contamination in NiTi(-X) Alloy on Vacuum Induction Melting in Various Crucibles

Crucible material	Duration of melting, min	Interaction layer thickness, μm	Impurity content, wt.%	Reference
SiC	90	Not reported	C: 1.129	Chapin and Friske 1954b, Sadrnezhaad and Raz 2005
Al$_2$O$_3$	90	Not reported	O: 1.6	Chapin and Friske 1954b, Sadrnezhaad and Raz 2005
ZrO$_2$	90	Not reported	O: 1.046	Chapin and Friske 1954b, Sadrnezhaad and Raz 2005
CaZrO$_3$	5	30		Li et al. 2010
BaZrO$_3$	2	20	O: 0.065–0.09	Chen et al. 2018b
High-density graphite	0.2–49	2–5	O: 0.02–0.082C: 0.017–0.25	Frenzel et al. 2004, Kim et al. 2001, Kramer 2009, Otubo et al. 2006, Zhang et al. 2005, 2006

Source: Fashu et al. 2020.

ingested in NiTi alloys reacts with titanium and forms TiC, the Ni 'content' increases and causes a shift of phase transition temperatures. A shift of 0.1 at.% in Ni content causes a 10 K difference in phase transition temperatures. Oxygen and nitrogen are also detrimental impurities. The impurity ingestion in NiTi alloys on melting in various crucibles is summarized in Table 4.7 (Fashu et al. 2020).

As well as the right composition, a good chemical homogeneity is also needed in the alloy ingot. Production of SMA components involves hot working of the homogeneous alloy and machining to the final shape (Frenzel et al. 2004).

Both BaZrO$_3$ and high-density graphite are suitable crucible materials for melting the NiTi (-X) alloys (Fashu et al. 2020). In addition, certain on-the-spot tricks are used, like lining graphite crucibles with titanium pieces while melting nickel–titanium alloys. The arrangement of the charge in the crucible is as shown in Figure 4.23 (Frenzel et al. 2004; Zhang et al. 2005). This arrangement ensures that the protective TiC layer forms quickly and early on during melting on the graphite crucible (Zhang et al. 2005), and right from the first melting, the direct contact between Ni and graphite is avoided. The net result is stopping the carbon transport from the graphite to the melt and minimum carbon contamination. The maximum acceptable oxygen and carbon content of NiTi alloys should not exceed 0.05 wt.% (Nayan et al. 2007). Table 4.7 shows the crucible–melt interaction data for NiTi-based SMAs melted using different crucibles. Melts in ceramic oxide crucibles are not pure enough (Fashu et al. 2020).

Titanium melts at 1670°C, the melting point of nickel is 1455°C, and a NiTi melt with 50.7 at.% Ni solidifies at 1310°C. Frenzel et al. (2004) prepared a NiTi alloy

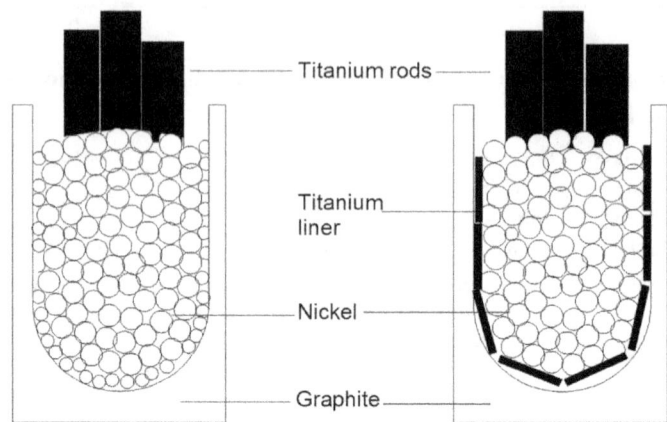

FIGURE 4.23   Arrangement of charge in graphite crucible for NiTi melting (From Frenzel et al. 2014).

from Ni pellets and Ti rods by VIM in a graphite crucible, which was also the susceptor. The batch size was 1 kg and the target composition Ni-50.7–Ti-49.3 (at .%). Thick discs (2 mm) were cut from the rod and used for cladding the crucible from inside in some experiments.

In melting, the target temperature of 1550°C was reached within 8 min, and melting of Ni and subsequent dissolution of Ti in the Ni melt occurred within the next 1 min. The melt was held for 10s more to ensure sufficient mixing. Frenzel et al. (2004) noted that in both the loading scenarios, nickel melted first, and then, the Ti rods slowly sank and dissolved into the melt (in approximately 1 min). The melt was then poured into a preheated (550°C) Y-shaped steelmould with an Y_2O_3 coating. Figure 4.24 shows carbon and oxygen concentrations in NiTi ingots processed in crucibles that were used for the first (1) and second (2) time.

The first set in Figure 4.24 refers to the ingot that was produced using a crucible filling arrangement where Ni was in direct contact with the graphite. Here, when the crucible was used for the first time, the carbon content of the NiTi ingot is high. In the second melting in the same crucible, the residual element contents come down to lower than benchmark levels. In the second set, titanium discs are in direct contact with graphite and not nickel. The carbon (and oxygen) contents in both the ingots are far below the benchmark levels right from the first melting. The carbon–titanium reaction is inhibited by many factors: (i) the solubility of carbon in solid Ti is very small, (ii) the direct physical contact between the flat titanium plate lining and the curved inner surface of the graphite crucible cannot be good, and (iii) the reaction is additionally hindered by diffusion barriers formed by the initial interaction. One graphite crucible was repeatedly used by Frenzel et al. (2004) for as many as 10 batches, and the carbon and oxygen contamination levels remained quite reproducible and significantly below the accepted industrial standard. Irrespective of the way the components are arranged in the crucible, the carbon ingestion into the melt in the second melting in the same crucible is lower. This is attributed by Frenzel et al. (2004) to the formation of a TiC layer in

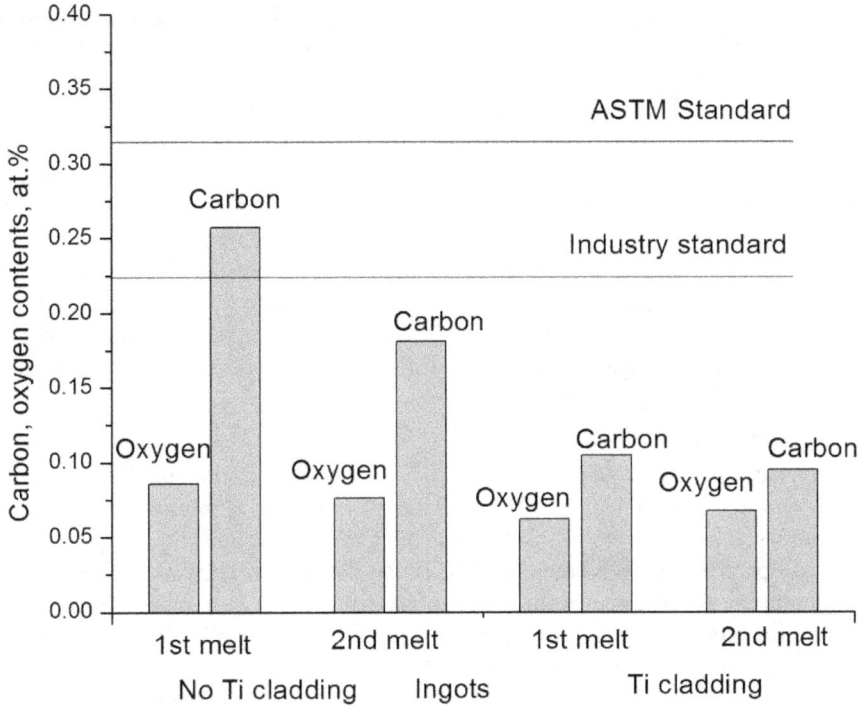

FIGURE 4.24 Carbon and oxygen contents in NiTi melted in graphite crucibles (From Frenzel et al. 2004).

the inner surface of the crucible by the time the first melt gets over. SEM micrographs revealed the presence of the reaction layer after the first melt. Significantly, the thickness of the reaction layer does not change much with the number of melts completed in the crucible. In a campaign of multiple melting, the special charge arrangement needs to be done for only the very first melting. All subsequent meltings can be done following the easier arrangement (no cladding).

Elahinia et al. (2012) also used a graphite crucible in VIM but did not allow the melt temperature to exceed 1450°C. This was 100°C lower than the temperature used by Frenzel et al. (2004). The carbon content in the ingot was controlled to between 200 and 500 ppm, and at this level, carbon does not affect the shape memory characteristics of the alloy. Elahinia et al. (2012) used the procedure and arrangement outlined by Frenzel et al. (2004) but doubled the batch size to 1 kg. The carbon content in final products ranges from 0.04% to 0.06% (Otubo et al. 2003b). VIM also ensures melt homogeneity by the strong electromagnetic stirring.

Otubo et al. (2006), in a classic investigation, directly addressed the effects of graphite quality and batch size of the NiTi melt on achieving the residual carbon content acceptable for commercial NiTi SMA products. The ingot produced in a high-porosity graphite crucible from a charge of Ti plates intercalated with electrolytic Ni plates presented the highest carbon contamination of 2440 ppm (ppm are in wt.%). Changing to a low-porosity crucible, the contamination was reduced by a factor of two (1000–1880 ppm). This decrease is due to a decrease of actual contact area due to the change from a high-porosity graphite crucible to a low-porosity crucible of the same apparent contact area. As the melting batch size was scaled up, and in the process, the specific contact area (area of contact with the graphite per unit weight of the melt) decreased, the average carbon content was reduced by another factor of two to reach around 600 ppm. The oxygen content generally remained between 400 and 900 ppm. This reduction in carbon contamination is the consequence of the change in the ratio between apparent contact area and liquid volume, which decreased from 0.68 to 0.43, representing a reduction of 37%, while the ingot mass increased from 1.5 kg in the second stage to 4.3 kg in the third stage, that is, almost three times.

4.25.9.1   Barium Zirconate (Calcia-Doped) Crucibles

Barium zirconate shows good corrosion resistance to TiNi, TiFe and TiAl melts, but it still exhibited insufficient stability against alloys rich in titanium. This limitation was addressed by Chen et al. (2018a) by using CaO-doped $BaZrO_3$ crucibles, which were less corroded by Ti2Ni melt than the $BaZrO_3$ crucible. The CaO-doped $BaZrO_3$ refractory may be potentially the more suitable refractory for induction melting of the titanium alloys. Chen et al. (2017) also carried out similar experiments with yttria-doped $BaZrO_3$ to melt TiNi and Ti2Ni alloys. Doping with Y_2O_3 decreases the sinterability of the $BaZrO_3$ crucible, but the crucible becomes more resistant to melt attack. The oxygen content of the TiNi alloy increased from 0.045% in the charge to 0.058% in the plain crucible but only to 0.051% in the doped crucible. The corresponding values in the case of Ti2Ni were 0.36% and 0.09% oxygen. The contamination by barium and zirconium was also similarly much less in the alloy melted in the doped crucible compared with the one melted in the plain barium zirconate crucible. The extent of erosion of yttria-doped $BaZrO_3$ crucible by both the melts is lower than that of the plain $BaZrO_3$ crucible. The yttria-doped $BaZrO_3$ appeared more suitable for induction melting of titanium alloys (Chen et al. 2017).

4.25.9.2   VAR and EBM

SMA components have been commercially produced by argon arc melting of pure components to get a homogeneous alloy, followed by a hot working and machining to the final shape (Frenzel et al. 2004).

The use of the EBM process to produce NiTi SMA in practice was begun by Otubo et al. (2003a, 2003b) in Japan. Nickel has a higher vapour pressure than Ti and can evaporate during EBM. The nickel loss, which was the initial concern in using EBM, can be overcome by controlling the EB power and the initial composition of the raw materials. Otubo et al. (2004) have shown that composition control is possible in NiTi alloys as long as certain precautions are taken during the feed material preparation. EBM product has a lower carbon content, ranging from 0.007% to 0.016% compared with 0.04% to 0.06% in the NiTi alloy commercially made by VIM (Otubo et al. 2003b). Oxygen impurity is also lower (Otubo et al. 2004).

4.25.10   HYDROGEN STORAGE ALLOYS

As a class of titanium-based functional materials, hydrogen storage alloys, like SMA, also have a critical requirement for composition control. They are very sensitive to contamination by non-metallic impurities, deviations from the target composition and inhomogeneity.

Fashu et al. (2020) suggest, based on published information, the use of crucibles made of CaO, $BaZrO_3$ or $CaZrO_3$ for melting TiFe-based alloys. The sensitivity of CaO refractories towards water vapour can be largely resolved by doping with Y_2O_3, which results in the formation of metastable CaY_2O_4 phase on grain boundaries. One technique used to manage contamination issues in alloy melting is to minimize the contact time between the crucible and the reactive components in the melt. The less reactive components are pre-alloyed, e.g. Mn with FeV, before co-melting them with more reactive (rare earth metal deoxidizer) and more refractory (titanium, zirconium, chromium) components. In certain practices, Ti- and Zr-containing AB2-type hydrogen storage alloys are pre-melted in large-capacity (>1000 kg/load) crucible-free plasma heated skull furnaces (Friedrich 1994).

TiFe alloy melted in a $BaZrO_3$ crucible had better hydrogen storage properties than one melted in a graphite crucible (Li et al. 2015). The inertness of $BaZrO_3$ was enhanced after doping with CaO (Chen et al. 2018a) and that of $CaZrO_3$ after adding Al_2O_3 (Lu et al. 2017, 2019). This was attributed to altered wetting behaviour of these ceramic crucibles. $CaZrO_3$ doped with Al_2O_3, Y_2O_3 and MgO has been shown to possess higher chemical stability and good tolerance against thermal shock.

The melting of TiFe in a graphite crucible (Li et al. 2015) results in contamination of the ingot by carbon. During melting, TiC forms and sticks strongly to the crucible wall. On melting Ti- and Zr-based AB2-type alloys containing Mn, carbides of manganese form even at lower temperatures, below 1300°C. Severe interaction of the melt with the graphite can even cause the crucible to disintegrate.

Fashu et al. (2020) have observed a satisfactory performance of crucibles made of Al_2O_3, $Al_2O_3 \cdot SiO_2$ and low-density graphite lined with multilayer coatings where the external (facing the melt) layer used was Y_2O_3. In trial melts, the AB2-type alloy did not stick to the crucible even after solidification inside, and displayed hydrogen sorption properties similar to a reference sample made by arc melting.

4.26   SUMMARY

In this chapter, the status as regards the metal–crucible systems in some of the most important metal processing operations has been summarized. Few metals have been as closely dependent on availability of crucibles for their extraction as iron. The reduction process predisposes the as-produced iron, called the hot metal or pig iron, to be liberally contaminated with a variety of impurities. The metal is still usable as such for a few applications (after a certain amount of metallurgical tending), but as the impurities are progressively decreased both in number and in concentration, the variety of uses increases greatly, and the uses become more sophisticated. 'Clean steel', which has less than 50 ppm of residual impurities (PCSNHO), is the cleanest metal produced routinely under industrial conditions in tonnage quantities. The process metallurgy part of iron and steel is all about impurities control, i.e. removal of impurities from the metal and protection of the metal from re-ingestion of impurities during further processing to a finished product. Refractories (crucibles) play a direct role here. The refractories undergo exemplary physical, thermal and chemical stresses in the various equipment for the making of iron and steel, and these facilities have long been the test track for new and improved refractories, not only for ferrous metallurgy but for all non-ferrous metallurgical operations as well.

Almost all the world's aluminium is produced by electrolysis in Hall–Héroult cells. The issues of metal–crucible interaction in the Hall–Héroult cell are mainly concerned with the interaction of the molten metal and the bath components with the cathode and the refractory layer beneath the cathode. Even though the temperatures involved are relatively low, interactions are many, and intensity is high. In the design and construction of the cell, the objective is to have the 800–850°C isotherms within the main body of the upper, protective refractory layer (Siljan et al. 2002). The solidus of the advancing melt falls in this temperature window. The most important purpose of the refractory layer, i.e. as a barrier to bath intrusion and to avoid deterioration of the thermal insulation layers below, will then be served.

Key interactions to be considered in the cell involve the electrolyte, molten aluminium and sodium with the cell materials, particularly the cathode carbon block and the refractory beneath it.

Sodium is generated and also transported through the cathode carbon. As well as sodium and electrolyte, aluminium also ends up at the other side of the cathode due to damage to the cathode. The pathways created by the damage allow liquid aluminium to penetrate down to the top of the refractory lining. Then, a number of reactions occur between the penetrating materials and the refractory, which result in the formation of new phases, causing volume expansion and also viscosity changes in the melt. Typically, the barrier refractory layer is made of aluminosilicates (usually fireclay), which are prone to infiltration and attack by molten electrolyte components. Conditions are chosen such that the reaction among cryolite, sodium fluoride, alumina, silica and mullite lead to the formation of albite ($NaAlSi_3O_8$), which raises the viscosity so much that melt intrusion into the barrier refractory is halted. In practice, this (albite-rich melt) is achieved by selecting a refractory with silica content above 60% and by a controlled input of sodium compounds, e.g. with excess sodium fluoride.

As-reduced aluminium from the Hall–Héroult cell is purified further in post-reduction treatments by fluxing (and skimming) and degassing before casting the 99.8–99.85% Al into moulds. The general requirements for refractories in aluminium melting furnaces are porosity <15%, average pore size <5 μm and low wetting by Al. The use of anti-wetting additions to refractories in contact with molten aluminium is a popular approach. Many different materials have been used for this purpose; the details remain proprietary, and so do the exact mechanism(s) of their action.

The furnaces used for producing molten copper from concentrates and scrap face stiff challenges to the refractory life in the form of highly aggressive slags, mechanical stresses, batch operation and higher operating temperatures. Mag–chrome is the refractory that serves well here, with the composition and type tweaked depending on the zone of its use. When everything appeared to be settling down, the use of 'mag–chrome' refractories began to be scrutinized and questioned as being a source of a possible environmental villain, hexavalent chromium. In a way, the search for a 'more' suitable refractory has begun all over again. Mag–alumina instead of mag–chrome is a possibility in almost every copper production application. Periclase–spinel or spinel–periclase might play a wider role. Doloma and mag–doloma might be an option for refining furnaces, where highly basic slags and highly oxidized copper make the replacement of mag–chrome a work in progress.

Due to their notoriously reactive nature and the need for elevated temperatures for their production and processing, the task of handling almost every member of the less common metals group has been challenging. Metallothermic reductions are generally exothermic reactions and have been extensively used for the preparation of less common metals.

For a host of technical reasons, the compound chosen is either an oxide or a fluoride or a chloride, and the reducing agent is aluminium, calcium, magnesium, sodium, potassium or lithium. Reduction using a metal invariably needs a high temperature, attained either by external heating or by the exothermicity of the reaction or by a combination of both. An inert containment is a prerequisite to protect the compound intermediate, the reducing agent and the as-reduced metal from contamination. The seriousness of a reactive melt increases first of all with its reactivity and also when the melt forms at a higher temperature and in larger quantities and persists for longer durations. A bomb reactor, which is a refractory-lined steel vessel that is usually closed during the course of a reduction, is often used. The refractory lining is principally used to contain the melt while it forms and separates into two layers – metallic melt as the bottom layer and slag as the top layer. The linings used include dolomite, magnesia, alumina, and even calcium fluoride and magnesium fluoride when these fluorides form as a product of metallothermic reduction. In another large group of reductions, unalloyed tantalum crucibles are the containers (for rare earth metals) in reactions conducted in vacuum induction furnaces. In relatively lower-temperature reductions, molybdenum or titanium containers are frequently used, and the reactor inside may also be lined with molybdenum or tantalum. In certain sophisticated reduction arrangements, water-cooled copper vessels are used as crucibles, and it is the thin layer of the metal being processed frozen on the water-cooled vessel that is the de facto container.

Fused salt electrolysis methods score over metallothermic reduction as superior processes for certain rare metal reductions. The cells are built using graphite, molybdenum and tungsten as materials that contact the electrolyte and the reduced metal. A technically sophisticated method, pioneered by the Reno group, is to collect the reduced metal in a frozen electrolyte skull. Even yttrium (melting point 1526°C) was electrowon in a fluoride electrolyte cell operated at 1700°C.

The melting and casting of titanium brings forth every conceivable challenge that can exist in the melt processing of a reactive metal. Perhaps, only vanadium surpasses titanium in being the unsolved problem in this context. The critical process is alloy melting. The crucible–melt interactions in a VIM operation depend on the titanium content of the alloy as well as the temperature needed for its melting. In practice, TiAl(γ) alloys can be melted at temperatures between 1550 and 1750°C, while Ti6Al4V alloys require 1660–1800°C due to the high Ti content. Until now, no refractory material has been found to be absolutely inert against these alloys, and some interactions between melt and crucible and even mould materials have always occurred during melting and casting. Cold crucible methods have been used to keep contamination levels low, but they have other issues, such as batch size and superheat. Among the large number of crucible options examined by numerous investigators, yttria, calcia, aluminium nitride, graphite, barium zirconate and calcium zirconate have been the most promising. The usage may involve a monolithic crucible or a coating. Impurities and porosity need to be closely controlled to derive the maximum benefit of chemical stability provided by these crucible materials.

In this chapter, building on the matter presented in the earlier chapters, all major and tangible metal–crucible interaction scenarios have been examined. It is true that in the field of metals processing, there are literally hundreds of interaction situations, each unique in its own way and depending on many parameters, only some of which are quantifiable or quantified. It will be highly desirable if a ready reckoner can be made to accurately predict the performance expected of a given melt–crucible combination. But, such an effort may lead to a path going nowhere, because the drivers of each published interaction are not available for complete and accurate listing. The way to go is, therefore, armed with the information available so far, to try to solve the emerging problems by analogy and experimentation. This volume is created with the intention of helping the reader to do this. The quest is on.

References

Adabifiroozjaei, E., Koshy, P., and Sorrell, C.C. (2015). Assessment of non-wetting materials for use in refractories for aluminium melting furnaces. *Journal of the Australian Ceramic Society*, 51(1): 139–145.

Afshar, S. and Allaire, C. (1996). The corrosion of refractories by molten aluminum. *JOM: The Journal of The Minerals, Metals & Materials Society*, 48(5): 23–27.

Afshar, S. and Allaire, C. (2001). Furnaces: Improving low cement castables by non-wetting additives. *JOM:The Journal of The Minerals, Metals & Materials Society*, 53(8): 24–27.

Aguilar-Santillan, J. (2008). Wetting of Al_2O_3 by molten aluminum: The influence of $BaSO_4$ additions. *Journal of Nanomaterials*, Vol. 2008, Article ID 629185, 12 pp.

Alangi, N., Mukherjee, J., Anupama, P., Verma, M.K., Chakravarthy, Y., Padmanabhan, P.V.A., Das, A.K., and Gantayet, L.M. (2011). Liquid uranium corrosion studies of protective yttria coatings on tantalum substrate. *Journal of Nuclear Materials*, 410(1–3): 39–45.

Allaire, C. (1992). Refractories for the lining of holding and melting furnaces. In: M.M. Avedesian et al. (Ed.), *Advanced in production and fabrication of light metals and metal matrix composites*. CIM, Montreal, Canada, pp. 163–174.

Allaire, C. (2000). Refractories for molten aluminium confinement. *Canadian Ceramics*, 69(1): 14–21.

Allen, N.P., Kubaschewski, O., and Von Goldbeck, O. (1951). The free energy diagram of the vanadium-oxygen System. *Journal of the Electrochemical Society*, 98(11): 417–424.

Baldwin, W.J. (1948). Zircon and zirconia refractories. *Chemical Engineering Progress*, 44: 875.

Barbosa, J. and Ribeiro, C.S. (2000). Influence of crucible material on the level of contamination in TiAl using induction melting. *International Journal of Cast Metals Research*, 12(5): 293–301.

Barbosa, J., Ribeiro, C.S., and Monteiro, A.C. (2003). Processing of γ- TiAl by ceramic crucible induction melting, and pouring in ceramic shells. *Materials Science Forum*, 426–432: 1933–1938.

Barbosa, J., Ribeiro, C.S., Teodoro, O.M.N.D., and Monteiro, A.C. (2005). Evaluation of Y2O3 front layer of ceramic crucibles for vacuum induction melting of TiAl based alloys, EPD Congress 2005. In: *Proceedings of sessions and symposia TMS annual meeting*, TMS, Warrendale, PA.

Barbosa, J., Puga, H., Ribeiro, C.S., Teodoro, O.M.N.D., and Monteiro, A.C. (2006). Characterisation of metal/mould interface on investment casting of γ-TiAl. *International Journal of Cast Metals Research*, 19(6): 331–338.

Barbosa, J., Ribeiro, C.S., and Monteiro, A.C. (2007). Influence of superheating on casting of γ- TiAl. *Intermetallics*, 15(7): 945–955.

Bayley, J. and Eckstein, K. (2006). Roman and medieval litharge cakes: Structure and composition. In: J. Pérez-Arantegui (Ed.), *Proceedings of the 34th international symposium on archaeometry, Institución Fernando el Católito*. CSIC, Zaragoza, pp. 145–153.

Beard, A.P. and Crooks,D.D. (1954). Kilogram scale reductions of vanadium pentoxide to vanadium metal, *J. Electrochem. Soc.*, 101(12): 597–600.

Beaudry, B.J. and Gschneidner, Jr., K.A. (1978). Preparation and basic properties of rare earth metals. In: K.A. Gschneidner, Jr. and L. Eyring (Eds.), *Handbook on the physics and chemistry of rare earths* (Vol. 1). North Holland Publishing Company, Amsterdam, pp. 173–232.

Biswas, S. and Sarkar, D. (2020). *Introduction to refractories for iron and steelmaking*, Springer International Publishing, Switzerland.

Block, F.E. (1960). *Preparation of high-purity yttrium by metallic reduction of yttrium trichloride, report of Investigations 5588.* Bureau of Mines, U.S. Department of the Interior, Washington, DC.

Bonadia, P., Braulio, M., CAllo, J.B., and Pandolfelli, V.C. (2006). Refractory selection for long-distance molten-aluminium delivery. *American Ceramic Society Bulletin*, 85(8): 9301–9309.

Booth, S.H. (1940). Use of refractories in melting copper and copper alloys in an Ajax-Wyatt induction furnace. *Bulletin of the American Ceramic Society*, 19: 171.

Bowen, N.L. and Greig, J.W. (1924). The system: Al_2O_3–SiO_2. *Journal of the American Ceramic Society*, 7(4): 238–254.

Brace, P.H. (1948). Reactions of molten titanium with certain refractory oxides. *Journal of the Electrochemical Society*, 94(4): 170–176.

Braulio, M.A.L., Martinez, A.T., Luz, A.P., Liebske, C., and Pandolfelli, V.C. (2011). Basic slag attack of spinel-containing refractory castables. *Ceramics International*, 37(6): 1935–1945.

Bravo, I.M.V., PINA, I.P., and Ransanz, G.R. (2017). Electrofused magnesio-chromite: Complex refractory material. *Macla: Revista de la Sociedad Española de Mineralogía*, 22: 89–90.

Brewer, L., Bromley, L.A., Gilles, P.W., and Lofgren, N.L. (1948). *The preparation and tests of refractory sulfide crucibles* (September 7), AECD-2253.

Bridges, P.J. and Hauzeur, F. (1991). Advances in the technology of titanium castings. *Cast Metals*, 4(3): 152–154.

Brondyke, K.J. (1953). Effect of molten aluminum on alumina-silica refractories. *Journal of the American Ceramic Society*, 36(5): 171–174.

Brook, R.J. (Ed.). (1991). *Concise encyclopedia of advanced ceramic materials*. Pergamon Press, Oxford.

Buhr, A., Bruckhausen, R., and Fahndrich, R. (2016). The steel industry in Germany–Trends in clean steel technology and refractory engineering. *Refractories World Forum*, 8(1): 57–63.

Bullock, E., Brunetaud, R., Condé, J.F., Keown, S.R., and Pugh, S.F. (Eds.). (1989). *Research and development of high temperature materials for industry*. Elsevier Applied Science, London.

Bunch, B.H. and Hellemans, A. (2004). *The history of science and technology: A browser's guide to the great discoveries, inventions, and the people who made them, from the dawn of time to today*. Houghton Mifflin Company, Boston, MA.

Calverley, A. and Lever, R.F. (1957). The floating-zone melting of refractory metals by electron bombardment. *Journal of Scientific Instruments*, 34(4): 142–147.

Calvert, E.D. (1981). *An investment mold for titanium casting, report of investigations 8541.* Bureau of Mines, U.S. Department of the Interior, Washington, DC.

Camel, D., Drevet, B., Brizé, V., Disdier, F., Cierniak, E., and Eustathopoulos, N. (2017). The crucible/silicon interface in directional solidification of photovoltaic silicon. *Acta Materialia*, 129: 415–427.

Canterford, J.H. (1985). Magnesia—An important industrial mineral: A review of processing options and uses. *Mineral Processing and Extractive Metallurgy Review*, 2(1–2): 57–104.

Carlson, O.N., Burkholder, H.R., Martsching, G.A., and Schmidt, F.A. (1981). Preparation of high purity vanadium. In: H.Y. Sohn, O.N. Carlson, and J.T. Smith (Eds.), *Extractive metallurgy of refractory metals—110th annual meeting, The Metallurgical Society of AIME.* Warrandale, PA, pp.191–203.

Carlson, O.N., Schmidt, F.A., and Krupp, W.E. (1966). A process for preparing high—Purity vanadium. *JOM:The Journal of The Minerals, Metals & Materials Society*, 18(3): 320–323.

Cegan, T. Szurman, I., Kursa, M., Holešinský, J., and Vontorová, J. (2016). Preparation of TiAl-based alloys by induction melting in graphite crucibles. *Kovove Materialy*, 53(2): 69–78.

Chakravarthy, Y., Bhandari, S., Chaturvedi, V., Pragatheeswaran, A., Nagraj, A., Thiyagarajan, T.K., Ananthapadmanaban, P.V., and Das, A.K. (2015). Plasma spray deposition of yttrium oxide on graphite, coating characterization and interaction with molten uranium. *Journal of the European Ceramic Society*, 35(2): 787–794.

Chang, Y.W. and Lin, C.C. (2010). Compositional dependence of phase formation mechanisms at the interface between titanium and calcia-stabilized zirconia at 1550°C. *Journal of the American Ceramic Society*, 93(11): 3893–3901.

Chapin, E.J., and Friske, W.H. (1954a). *A Metallurgical Evaluation of Refractory Compounds for Containing Molten Titanium, Part I-Oxides (No. NRL-4447)*. Naval Research Lab, Washington, DC.

Chapin, E.J. and Friske, W.H. (1954b). *A metallurgical evaluation of refractory compounds for containing molten titanium, part II-carbon, graphite, and carbides (No. NRL-4467)*. Naval Research Lab, Washington, DC.

Chapin, E.J. and Friske, W.H. (1955). *A metallurgical evaluation of refractory compounds for containing molten titanium. Part III-borides and sulfides (No. NRL-4478)*. Naval Research Lab, Washington, DC.

Chen, G., Gao, P., Kang, J., Li, B., Ali, W., Qin, Z., Lu, X., and Li, C. (2017). Improved stability of $BaZrO_3$ refractory with Y_2O_3 additive and its interaction with titanium melts. *Journal of Alloys and Compounds*, 726: 403–409.

Chen, G., Kang, J., Gao, P., Ali, W., Qin, Z., Lu, X., and Li, C. (2018a). Effect of CaO additive on the interfacial reaction between the $BaZrO_3$ refractory and titanium enrichment melt. In: Kim, H., Wesstrom, B., Alam, S., Ouchi, T., Azimi, G., Neelameggham, N. R., Wang,S., Guan, X. (Eds.) *Rare Metal Technology*. Springer, Cham, pp. 235–244.

Chen, G., Kang, J., Gao, P., Qin, Z., Lu, X., and Li, C. (2018b). Dissolution of $BaZrO_3$ refractory in titanium melt. *International Journal of Applied Ceramic Technology*, 15(6): 1459–1466.

Chen, G., Kang, J., Lan, B., Gao, P., Lu, X., and Li, C. (2018c). Evaluation of Ca-doped BaZrO3 as the crucible refractory for melting TiAl alloys. *Ceramics International*, 44(11): 12627–12633.

Chen, G., Lan, B., Xiong, F., Gao, P., Zhang, H., Lu, X., and Li, C. (2019). Pilot-scale experimental evaluation of induction melting of Ti-46Al-8Nb alloy in the fused $BaZrO_3$ crucible. *Vacuum*, 159: 293–298.

Chindgren, C.J., Bauerle, L.C., and Rosenbaum, J.B. (1963). *Modifications in bomb reduction of vanadium oxide, report of investigations 6284*. Bureau of Mines, U.S. Department of the Interior, Washington, DC.

Clites, P.G. and Calvert, E.D. (1961). Laboratory casting high-temperature metals. *JOM: The Journal of The Minerals, Metals & Materials Society*, 13(2): 136–138.

Cooper, A.R. and Kingery, W.D. (1959). Corrosion of refractories by liquid slags and glasses. In: W.D. Kingery (Ed.), *Kinetics of high-temperature processes*. John Wiley and Sons, New York, pp. 85–92.

Cooper, D.E., Snow, K.S., Fulton, J.C., and Tommaney, J.W. (1965). Feasibility experiments of vacuum arc remelting of 28 in.-diameter ingots in three-electrode AC furnace. *JOM: The Journal of The Minerals, Metals & Materials Society*, 17(12): 1368–1373.

Croat, J.J. (1969). *The preparation of high purity dysprosium, holmium and erbium by the lithium reduction of their trichloride salts, No. IS-T-346, Ames Laboratory, ERDA, Iowa State University, Ames, IA*. National Technical Information Service, Springfield, VA.

Cui, R., Gao, M., Zhang, H., and Gong, S. (2010). Interactions between TiAl alloys and yttria refractory material in casting process. *Journal of Materials Processing Technology*, 210(9): 1190–1196.

Cui, R.J., Tang, X.X., Gao, M., Zhang, H., Gong, S.K. (2012). Microstructure and composition of cast Ti-47Al-2Cr-2Nb alloys produced by yttria crucibles, *Mater. Sci. Eng. A* 541: 14–21.

Daane, A.H. and Spedding, F.H. (1953). Preparation of yttrium and some heavy rare earth metals, *J. Electrochem. Soc.*, 100: 442–444.

Daane, A.H., Dennison, D.H. and Spedding, F.H. (1953). The preparation of samarium and ytterbium metals, *J. Am. Chem. Soc.*, 75: 2272–2273.

Daane, A.H. (1961a). Metallothermic preparation of rare earth metals. In: F.H. Spedding and A.H. Daane (Eds.), *The rare earths*. Wiley, New York, pp. 102–112.

Daane, A.H. (1961b). Yttrium. In: C.A. Hampel (Ed.), *Rare metals handbook*. Reinhold, London, pp. 653–666.

Daane, A.H. and Spedding, F.H. (1953). Preparation of yttrium and some heavy rare earth metals. *Journal of the Electrochemical Society*, 100(10): 442–444.

Darling, A.S. (1990). Non-ferrous metals. In: I. McNeil (Ed.), *An encyclopaedia of the history of technology*. Routledge, London, pp. 47–145.

Dennison, D.H., Tschetter, M.J., and Gschneidner, Jr., K.A. (1966a). The solubility of tantalum in eight liquid rare earth metals. *Journal of the Less Common Metals*, 10: 108–115.

Dennison, D.H., Tschetter, M.J., and Gschneidner, Jr., K.A. (1966b). The solubility of Ta and W in liquid rare earth metals. *Journal of the Less Common Metals*, 11: 423–435.

Dunsing, T.W. (2002). Tips on good refractory practice for aluminum-melting furnaces. *Industrial Heating*, 69(4): 27–29.

Durov, A.V., Naidich, Y.V., and Kostyuk, B.D. (2005). Investigation of interaction of metal melts and zirconia. *Journal of Materials Science*, 40(9–10): 2173–2178.

Eastman, E.D., Brewer, L., Bromley, L.A., Gilles, P.W., and Lofgren, N.I. (1951). Preparation and tests of refractory sulfide crucibles. *Journal of the American Ceramic Society*, 34(4): 128–136.

Eastwood, L.W. and Craighead, C.M. (1950). *Refractories for melting titanium part I (No. A F-TR-6218 (pt. 1))*. Battelle Memorial Inst., National Technical Information Service, Springfield, VA.

Eatesami, D., Hadavi, M.M., and Habibollahzade, A. (2009). Melting of γ-TiAl in the alumina crucible. *Russian Journal of Non-Ferrous Metals*, 50(4): 363–367.

Economos, G. and Kingery, W.D. (1953). Metal-ceramic interactions: II, Metal-oxide interfacial reactions at elevated temperatures. *Journal of the American Ceramic Society*, 36(12): 403–409.

Elahinia, M.H., Hashemi, M., Tabesh, M., and Bhaduri, S.B. (2012). Manufacturing and processing of NiTi implants: A review. *Progress in Materials Science*, 57(5): 911–946.

Ellingham, H.J.T. (1944). Reducibility of oxides and sulphides in metallurgical processes. *Journal of the Society of Chemical Industry*, 63(5): 125–160.

Emeis, R. (1954). Tiegelfreies Ziehen von Silicium-Einkristallen. *Zeitschrift für Naturforschung A*, 9(1): 67–68.

Engel, R. (2015). Refractory considerations. *Aluminium International Today*, 27(3): 23.

Erb, A., Walker, E., and Flükiger, R. (1995). BaZrO$_3$: The solution for the crucible corrosion problem during the single crystal growth of high-Tc superconductors REBa$_2$Cu$_3$O$_{7-\delta}$; RE = Y, Pr, *Physica C: Superconductivity*, 245(3–4): 245–251.

Fadeev, A.V., Bazhenov, V.E., and Koltygin, A.V. (2015). Improvement in the casting technology of blades for aviation gas-turbine engines made of TNM-B1 titanium aluminide alloy produced by induction crucible melting. *Russian Journal of Non-Ferrous Metals*, 56(1): 26–32.

Fahey, N.P., Swinbourne, D.R., Yan, S., and Osborne, J.M. (2004). The solubility of Cr_2O_3 in calcium ferrite slags at 1573 K. *Metallurgical and Materials Transactions B*, 35(2): 197–202.

Fashu, S., Lototskyy, M., Davids, M.W., Pickering, L., Linkov, V., Tai, S., Renheng, T., Fangming, X., Fursikov, P.V., and Tarasov, B.P. (2020). A review on crucibles for induction melting of titanium alloys. *Materials and Design*, 186: 108295.

Field, T.E. (1947). Heat-cast refractory for magnesium melting furnaces, U. S. Pat. 2,416,472 (Feb. 25, 1947).

Finlay, G.R. and Fetterly, G. (1952). Boron nitride — An unusual refractory. *Journal of the American Ceramic Society*, 31: 141.

Forrester, R. (2019). *History of metallurgy* (2nd Ed.). SSRN Electronic Journal. doi:10.2139/ssrn.2864178.

Freitag, L., Schafföner, S., Lippert, N., Faßauer, C., Aneziris, C.G., Legner, C., and Klotz, U.E. (2017). Silica-free investment casting molds based on calcium zirconate. *Ceramics International*, 43(9): 6807–6814.

Frenzel, J., Zhang, Z., Neuking, K., and Eggeler, G. (2004). High quality vacuum induction melting of small quantities of NiTi shape memory alloys in graphite crucibles. *Journal of Alloys and Compounds*, 385(1–2): 214–223.

Friedrich, B. (1994). Large-scale production and quality assurance of hydrogen storage (battery) alloys. *Journal of Materials Engineering and Performance*, 3(1): 37–46.

Friedrich, B., Lochbichler, C., Brenk, J., Vonderstein, C., Schafföner, S., and Aneziris, C. (2016). Ceramic crucible melting of titanium alloys and intermetallics in VIM, Presentation, April 2016, Titanium Europe 2016, Paris. https://doi.org/10.13140/RG.2.1.3992.6161

Frueh, C., Poirier, D.R., Maguire, M.C., and Harding, R.A. (1996). Attempts to develop a ceramic mould for titanium casting—A review. *International Journal of Cast Metals Research*, 9(4): 233–239.

Frueh, C., Poirier, D.R., and Maguire, M.C. (1997). The effect of silica-containing binders on the titanium/face coat reaction. *Metallurgical and Materials Transactions B*, 28(5): 919–926.

Fruehan, R.J. (Ed.). (1998). *The making, shaping and treating of steel; Hubble, D.H., chapter 3 steel plant refractories; Hubble, D. H., Russell, R.O., Vernon, H.L. and Marr, R.J., chapter 4, steelmaking refractories* (11th Ed.). The AISE Steel Foundation, Pittsburgh, PA.

Gale, W.K.V. (1990). Ferrous metals. In: I. McNeil (Ed.), *An encyclopaedia of the history of technology*. Routledge, London, pp. 146–185.

Gao, M., Cui, R., Ma, L., Zhang, H., Tang, X., and Zhang, H. (2011). Physical erosion of yttria crucibles in Ti–54Al alloy casting process. *Journal of Materials Processing Technology*, 211(12): 2004–2011.

Gayler, M.L.V. (1930). The preparation and properties of chemically pure metals. *Metallwirtscbaft*, 9: 677.

Gingerich, K.A. (1968). Gaseous metal nitrides. II. The dissociation energy, heat of sublimation, and heat of formation of zirconium mononitride. *The Journal of Chemical Physics*, 49(1): 14–18.

Glazunov, S.G. (1970). Precision casting of titanium. In: R.I. Jaffee and N.E. Promisel (Eds.), *The science, technology and application of titanium*. Elsevier, Amsterdam, pp. 143–147.

Goldschmidt, H. and Vautin, C. (1898). Aluminium as a heating and reducing agent. *Journal of the Society of Chemical Industry*, 6(17): 543–545.

Gomes, F., Puga, H., Barbosa, J., and Ribeiro, C.S. (2011). Effect of melting pressure and superheating on chemical composition and contamination of yttria-coated ceramic crucible induction melted titanium alloys. *Journal of Materials Science*, 46(14): 4922–4936.

Gomes, F., Barbosa, J., and Ribeiro, C.S. (2013). Aluminium evaporation during ceramic crucible induction melting of titanium aluminides. *Materials Science Forum*, 730: 697–702.

Goto, K., Argent, B.B., and Lee, W.E. (1997). Corrosion of MgO—MgAl$_2$O$_4$ spinel refractory bricks by calcium aluminosilicate slag. *Journal of the American Ceramic Society*, 80(2): 461–471.

Gray, P.M.J. (1951). The production of pure cerium metal by electrolytic and thermal reduction processes. *Transactions of the Institution of Mining and Metallurgy*, 61: 141–170.

Griesenauer, N.M., Lyon, S.R., and Alexander, C.A. (1972). Vacuum induction melting of titanium. *Journal of Vacuum Science and Technology*, 9(6): 1351–1355.

Gupta, C.K. and Krishnamurthy, N. (1992). *Extractive metallurgy of vanadium*. Elsevier Science Publishers B. V., The Netherlands.

Hall, F.P. and Insley, H. (1933). *Phase diagrams for ceramists*. American Ceramic Society, Columbus, OH.

Hampel, C.A. (1961). *Rare metals handbook*. Reinhold Publishing Corporation, London.

Harding, R.A. and Wickins, M. (2003). Temperature measurements during induction skull melting of titanium aluminide. *Materials Science and Technology*, 19: 1235–1246.

Harding, R.A., Wickins, M., Wang, H., Djambazov, G., and Pericleous, K.A. (2011). Development of a turbulence-free casting technique for titanium aluminides. *Intermetallics*, 19(6): 805–813.

Harvey, F.A. (1945). Refractories. In: D.M. Liddell (Editor-in-Chief), *Handbook of nonferrous metallurgy*. McGraw-Hill, New York.

Henrie, T.A. and Morrice, E. (1966). V. A high temperature electrowinning cell for rare earths. *JOM: The Journal of The Minerals, Metals & Materials Society*, 18: 1207–1208.

Herget, C. (1985). Metallurgical ways to NdFeB alloys. Permanent magnets from cobalt-reduced NdFeB. (retroactive coverage). In: *Eighth international workshop on rare-earth magnets and their applications and the fourth international symposium on magnetic anisotropy and coercivity in rare earth--transition metal alloys*, 407–422.

Hirschhorn, I.S. (1968). Commercial production of rare earth metals by fused salt electrolysis. *JOM: The Journal of The Minerals, Metals & Materials Society*, 20(3): 19–221.

Hockaday, S.A.C., and Bisaka, K. (2010). Experience and results from running of a 1 kg Ti scale Kroll reactor. In: *Proceedings of the advanced metals initiative: Light metals conference 2010*. Southern African Institute of Mining and Metallurgy, Johannesburg. pp. 265–280.

Huffine, C.L. and Williams, J.M. (1961). Refining and purification of rare earth metals. In: F.H. Spedding and A.H. Daane (Eds.), *The rare earths*. Wiley, New York, pp. 145–162.

Jackson, C.M., Wagner, H.J., and Wasilewski, R.J. (1972). 55-Nitinol—the alloy with a memory: its physical metallurgy, properties, and applications Report NASA SP-5110.

Jain, S.C., Bose, D.K., and Gupta, C.K. (1971). *Pilot-plant production of capacitor-grade tantalum powder*. Bhabha Atomic Research Centre, Trombay, India.

Jamrack, W.D. (1963). *Rare metal extraction by chemical engineering techniques*. Pergamon, Oxford.

Johns, H.L., and King, A.G. (1970). Zirconia tailored for thermal shock resistance. *Ceram. Age*, 86(5): 29–31.

Jordan, L. and Swanger, W.H. (1930). Preparation of Pure Nickel, RP257. *Journal of Research of the National Bureau of Standards*, 5: 1291.

Kamat, G.R. and Gupta, C.K. (1971). Open aluminothermic reduction of columbium (Nb) pentoxide and purification of the reduced metal. *Metallurgical Transactions*, 2(10): 2817–2823.

Kamyshnykova, K. and Lapin, J. (2018). Vacuum induction melting and solidification of TiAl-based alloy in graphite crucibles. *Vacuum*, 154: 218–226.

Kaptay, G., Matsushita, T., Mukai, K., and Ohuchi, T. (2004). On different modifications of the capillary model of penetration of inert liquid metals into porous refractories and their connection to the pore size distribution of the refractories. *Metallurgical and Materials Transactions B*, 35(3): 471–486.

Kartavykh, A.V. and Cherdyntsev, V.V. (2008). Chemical compatibility of a TiAl-Nb melt with oxygen-free crucible ceramics made of aluminum nitride. *Russian Metallurgy (Metally)*, 2008(6): 491.

Kartavykh, A.V., Tcherdyntsev, V.V., and Zollinger, J. (2009). TiAl–Nb melt interaction with AlN refractory crucibles. *Materials Chemistry and Physics*, 116(1): 300–304.

Kartavykh, A.V., Tcherdyntsev, V.V., and Zollinger, J. (2010). TiAl–Nb melt interaction with pyrolytic boron nitride crucibles. *Materials Chemistry and Physics*, 119(3): 347–350.

Kaufmann, A.R. and Gordon, E. (1947). Vacuum melting and casting of beryllium. *Metal Progress*, 52: 387.

Keck, P.H. and Golay, M.J. (1953). Crystallization of silicon from a floating liquid zone. *Physical Review*, 89(6): 1297.

Kim, S., Hong, T., and Kim, Y. (2001). Evaluation of thermal stability of mold materials for magnesium investment casting. *Materials Transactions*, 42(3): 539–542.

Kim, J.H., Kim, H.T., Woo, Y.M., Kim, K.H., Lee, C.B., and Fielding, R.S. (2013). Interaction studies of ceramic vacuum plasma spraying for the melting crucible materials. *Nuclear Engineering and Technology*, 45(5): 683–688.

Klotz, U.E., Legner, C., Bulling, F., Freitag, L., Faßauer, C., Schafföner, S., and Aneziris, C.G. (2019). Investment casting of titanium alloys with calcium zirconate moulds and crucibles. *The International Journal of Advanced Manufacturing Technology*, 103(1) 343–353.

Knauft, R.W. (1943). Zircon refractories for aluminum melting furnaces. *Metals and Alloys*, 18: 1326.

Koger, J.W., Holcombe, C.E., and Banker, J.G. (1976). Coatings on graphite crucibles used in melting uranium. *Thin Solid Films*, 39: 297–303.

Komarek, K.L. and Silver, M. (1962). *Proc. IAEA Symp., Thermodynamics of Nuclear Materials*. Wien, 749.

Korgul, P., Wilson, D.R., and Lee, W.E. (1997). Microstructural analysis of corroded alumina-spinel castable refractories. *Journal of the European Ceramic Society*, 17(1): 77–84.

Kostov, A. and Friedrich, B. (2005). Selection of crucible oxides in molten titanium and titanium aluminum alloys by thermo-chemistry calculations. *Journal of Mining and Metallurgy, Section B*, 41(1): 113–125.

Kostov, A. and Friedrich, B. (2006). Predicting thermodynamic stability of crucible oxides in molten titanium and titanium alloys. *Computational Materials Science*, 38(2): 374–385.

Kramer, G.M. (2009). A comparison of chemistry and inclusion distribution and morphology versus melting method of NiTi alloys. *Journal of Materials Engineering and Performance*, 18(5): 479–483.

Kroll, W.J. (1940). The production of ductile titanium. *Transactions of the Electrochemical Society*, 78(1): 35–47.

Kroll, W.J. (1956). The pyrometallurgy of halides. *Metallurgical Reviews*, 1(1): 291–337.

Kroll, W.J. (1959). The present state of titanium extractive metallurgy. *Transactions of the American Institute of Mining and Metallurgical Engineers*, 215(4): 546–553.

Kroll, W.J. and Stephens, W.W. (1950). Pilot plants. Production of malleable zirconium. *Industrial and Engineering Chemistry*, 42(2): 395–398.

Kroll, W.J., Schlechten, A.W., and Yerkes, L.A. (1946). Ductile zirconium from zircon sand. *Transactions of the Electrochemical Society*, 89(1): 263.

Kroll, W.J., Schlechten, A.W., Carmody, W.R., Yerkes, L.A., Holmes, H.P., and Gilbert, H.L. (1947). Recent progress in the metallurgy of malleable zirconium. *Transactions of the Electrochemical Society*, 92(1): 99–113.

Kroll, W.J., Anderson, C.T., Holmes, H.P., Yerkes, L.A., and Gilbert, H.L. (1948). Large-scale laboratory production of ductile zirconium. *Journal of the Electrochemical Society*, 94(1): 1.

Kuang, R.J., Harding, P.A., and Campbell, J. (2000). Investigation into refractories as crucible and mould materials for melting and casting γ -TiAl alloys. *Materials Science and Technology*, 16(9): 1007–1016.

Kuang, R.J., Harding, P.A., and Campbell, J. (2001). A study of refractories as crucible and mould materials for melting and casting γ-TiAl alloys. *International Journal of Cast Metals Research*, 13(5): 277–292.

Kubaschewski, O. and Alcock, C.B. (1979). *Materials thermochemistry*. Pergamon, Oxford.

Kubaschewski, O. and Dench, W.A. (1954). The free-energy diagram of the system titanium oxygen. *Journal of the Institute of Metals*, 82: 87–91.

Kuznietz, M., Livne, Z., Cotler, C., and Erez, G. (1988). Diffusion of liquid uranium into foils of tantalum metal and tantalum-10 wt% tungsten alloy up to 1350°C. *Journal of Nuclear Materials*, 152(2–3): 235–245.

Lapin, J. (2009, May). TiAl-based alloys: Present status and future perspectives. In *Conference proceedings, METAL*, 19, No. 21.5.2009, 12 pp.

Lapin, J. and Klimová, A. (2019). Vacuum induction melting and casting of TiAl-based matrix in-situ composites reinforced by carbide particles using graphite crucibles and moulds. *Vacuum*, 169: 108930.

Lapin, J., Gabalcová, Z., and Pelachová, T. (2011). Effect of Y2O3 crucible on contamination of directionally solidified intermetallic Ti–46Al–8Nb alloy. *Intermetallics*, 19(3): 396–403.

Lee, W.E., and Zhang, S. (1999). Melt corrosion of oxide and oxide–carbonrefractories. *International Materials Reviews*, 44(3): 77–104.

Lee, W.E. and Moore, R.E. (1998). Evolution of in situ refractories in the 20th century. *Journal of the American Ceramic Society*, 81(6): 1385–1410.

Lee, W.E. and Rainforth, W.M. (1994). *Ceramic microstructures, property control by processing*. Chapman and Hall, London.

Lee, W.E. and Zhang, S. (2004, January). Direct and indirect slag corrosion of oxide and oxide-C refractories. In: *VII international conference on molten slags fluxes and salts*. The South African Institute of Mining and Metallurgy, Johannesburg, Symposium Series S 36, pp. 309–319.

Li, Z., Lee, W.E., and Zhang, S. (2007). Low-temperature synthesis of $CaZrO_3$ powder from molten salts. *Journal of the American Ceramic Society*, 90(2): 364–368.

Li, C.H., Gao, Y.H., Lu, X.G., Ding, W.Z., Ren, Z.M., and Deng, K. (2010). Interaction between the ceramic $CaZrO_3$ and the melt of titanium alloys. *Advances in Science and Technology*, 70: 136–140.

Li, M., Gehre, P., and Aneziris, C.G. (2013). Investigation of calcium zirconate ceramic synthesized by slip casting and calcination. *Journal of the European Ceramic Society*, 33(10): 2007–2012.

Li, C.H., He, J., Zhang, Z., Yang, B., Leng, H.Y., Lu, X.G., Li, Z.L., Wu, Z., and Wang, H.B. (2015). Preparation of TiFe based alloys melted by CaO crucible and its hydrogen storage properties. *Journal of Alloys and Compounds*, 618: 679–684.

Liu, K., Ma, Y.C., Gao, M., Rao, G.B., Li, Y.Y., Wei, K., Wu,X., and Loretto, M.H. (2005). Single step centrifugal casting TiAl automotive valves. *Intermetallics*, 13(9): 925–928.

Lin, K.F. and Lin, C.C. (1998). Interfacial reactions between zirconia and titanium. *Scripta Materialia*, 39(10): 1333–1338.

Lin, K.F. and Lin, C.C. (1999a). Interfacial reactions between Ti-6Al-4V alloy and zirconia mold during casting. *Journal of Materials Science*, 34(23): 5899–5906.

Lin, K.F. and Lin, C.C. (1999b). Transmission electron microscope investigation of the interface between titanium and zirconia. *Journal of the American Ceramic Society*, 82(11): 3179–3185.

Lochbichler, C. and Friedrich, B. (2007). Induction melting using refractaries and direct deoxidisation of Ti and TiAl Scrap. In: *European Metallurgical Conference*, EMC 2007, June 11-14, Düsseldorf, Germany.

Long, G. and Foster, L.M. (1959). Aluminum nitride, a refractory for aluminum to 2000°C. *Journal of the American Ceramic Society*, 42(2): 53–59.

López, V.H. and Kennedy, A.R. (2006). Flux-assisted wetting and spreading of Al on TiC. *Journal of Colloid and Interface Science*, 298(1): 356–362.

Lu, M.W., Lin, K.L., and Lin, C.C. (2017). *Effect of the alumina content on the interfacial reactions between titanium and calcia/zirconia/alumina composites, Contributed Papers from Materials Science and Technology 2017*, October 8–12, Pittsburgh.

Lu, M.W., Lin, K.L., and Lin, C.C. (2019). Investigation of the interactions between titanium and calcium zirconium oxide ($CaZrO_3$) ceramics modified with alumina. *Processing and Application of Ceramics*, 13(1): 79–88.

Lyon, S.R., Inouye, S., Alexander, C.A., and Niesz, D.E. (1973). The interaction of titanium with refractory oxides. In: *Titanium science and technology*. Springer, Boston, MA, pp. 271–284.

Malfliet, A., Lotfian, S., Scheunis, L., Petkov, V., Pandelaers, L., Jones, P.T., and Blanpain, B. (2014). Degradation mechanisms and use of refractory linings in copper production processes: A critical review. *Journal of the European Ceramic Society*, 34(3): 849–876.

Massalski, T.B. (1990a). *Binary alloy phase diagrams* (Vol. 1). American Society for Metals, Metals Park, OH.

Massalski, T.B. (1990b). *Binary alloy phase diagrams* (Vol. 2). American Society for Metals, Metals Park, OH.

McCollum, J.M. (1989). An improved 85% alumina phosphate-bonded refractory for molten aluminum contact. *Light Metal Age*, 47(11): 22–24.

Merrill, T.W. (1958). Ductile vanadium - A new engineering material. *JOM: The Journal of The Minerals, Metals & Materials Society*, 10: 618–621.

Mikami, H.M. and Martinet, J.R. (1979). Carbon magnesia bricks in electric arc furnaces *Refractories J*. 55(6): 25–32.

Moriarty, J.L. (1968). The industrial preparation of the rare earth metals by metallothermic reduction. *JOM: The Journal of The Minerals, Metals & Materials Society*, 20(11): 41–45.

Morrice, E. and Knickerbocker, R.G. (1961). Rare earth electrolytic metals. In: F.H. Spedding and A.H. Daane (Eds.), *The rare earths*. John Wiley, New York, pp. 126–144.

Morrice, E. and Wong, M.M. (1979). Fused salt electrowinning and electrorefining of rare earth and yttrium metals. *Minerals Science and Engineering*, 11(3): 125–136.

Morrice, E., Porter, B., Brown, E.A., Wyche, C., and Knickerbocker, R.G. (1961). *Electrowinning of cerium group and yttrium group metals, Report of Investigations 5868*. Bureau of Mines, U.S. Department of the Interior, Washington, DC.

Morrice, E., Shedd, E.S., and Henrie, T.A. (1968). *Direct electrolysis of rare earth oxides to metals and alloys in fluoride melts, Report of Investigations 7146*. Bureau of Mines, U.S. Department of the Interior, Washington, DC.

Morscheiser, J., Friedrich, B., and Lochbichler, C. (2008). Potential of ceramic crucibles for melting of titanium-alloys and gamma-titanium aluminide. 51 Internationales Feuerfestkolloquium, 15, 16.

Mukherjee, T.K. and Gupta, C.K. (1972). Open aluminothermic reduction of vanadium oxides. *Journal of the Less Common Metals*, 27(2): 251–254.

Nakamura, Y., Nakajima, H., Ishioka, S., and Koiwa, M. (1988). Effect of oxygen on diffusion of manganese in α-titanium. *Acta Metallurgica*, 36(10): 2787–2795.

Nandy, R.N. and Jogai, R.K. (2012). Selection of proper refractory materials for energy saving in aluminium melting and holding furnaces. *International Journal of Metallurgical Engineering*, 1(6): 117–121.

Nayan, N., Saikrishna, C.N., Ramaiah, K.V., Bhaumik, S.K., Nair, K.S., and Mittal, M.C. (2007). Vacuum induction melting of NiTi shape memory alloys in graphite crucible. *Materials Science and Engineerin A*, 465(1–2): 44–48.

Nolting, H.J., Simmons, C.R., and Klingenberg, J.J. (1960). Preparation and properties of high purity yttrium metal. *Journal of Inorganic and Nuclear Chemistry*, 14(3–4): 208–216.

Okabe, T.H., Suzuki, R.O., Oishi, T., and Ono, K. (1991). Thermodynamic properties of dilute titanium-oxygen solid solution in beta phase. *Materials Transactions, JIM*, 32(5): 485–488.

Orsten, S. and Oeters, F. (1986). Dissolution of carbon in liquid iron. In: *Process technology Proceedings*, 6: 143–155.

Otubo, J., Rigo, O.D., Neto, C.D.M., Kaufman, M.J., and Mei, P.R. (2003a). NiTi shape memory alloy ingot production by EBM. *Journal de Physique IV*, 112: 813–820.

Otubo, J., Rigo, O.D., Neto, C.D.M., Kaufman, M.J., and Mei, P.R. (2003b). Scale up of NiTi shape memory alloy production by EBM. In: *Journal de Physique IV (Proceedings)*, 112: 873–876. EDP Sciences.

Otubo, J., Rigo, O.D., Neto, C.D.M., Kaufman, M.J., and Mei, P.R. (2004). Low carbon content NiTi shape memory alloy produced by electron beam melting. *Materials Research*, 7(2): 263–267.

Otubo, J., Rigo, O.D., Neto, C.D.M., and Mei, P.R. (2006). The effects of vacuum induction melting and electron beam melting techniques on the purity of NiTi shape memory alloys. *Materials Science and Engineering A*, 438: 679–682.

Ou, J., Cockcroft, S.L., Maijer, D.M., Yao, L., Reilly, C., and Akhtar, A. (2015). An examination of the factors influencing the melting of solid titanium in liquid titanium. *International Journal of Heat and Mass Transfer*, 86: 221–233.

Pankratz, L.B., Stuve, J.M., and Gokcen, N.A. (1984). *Thermodynamic data for mineral technology*. Bureau of Mines Bulletin 677. U.S. Department of the Interior, Washington, DC.

Perfect, F.H. (1967). Metallothermic reduction of oxides in water-cooled furnaces. *Transactions of the Metallurgical Society of AIME*, 239(9): 1282–1286.

Perfect, F.H. (1981). Aluminothermic chromium and chromium alloys, low in nitrogen. *Metallurgical Transactions B*, 12(3): 611–613.

Phelps, G.W. and Wachtman, J.B. (2011). *Ullmann's encyclopedia of industrial chemistry*. Wiley-VCH Verrlag GmbH & Co, Weinheim.

Pidgeon, L.M. (1944). New methods for the production of magnesium. *The Canadian Institute of Mining and Metallurgy*, 47: 16–34.

Poirier, J. (2015). A review: Influence of refractories on steel quality. *Metallurgical Research and Technology*, 112(4): 410.

Rehren, T. (2003). Crucibles as reaction vessels in ancient metallurgy. In: P. Craddock and J. Lang (Ed.), *Mining and metal production through the ages*. British Museum Press, London, pp. 207–215.

Richardson, F.D. (1948). The thermodynamics of substances of interest in iron and steel making from 0 to 2400°C I-Oxides. *The Journal of the Iron and Steel Institute*, 160: 261–270.

Richardson, F.D., Jeffes, J.H.E., and Withers, G. (1950). The thermodynamics of substances of interest in iron and steel making-II-compounds between oxides. *The Journal of the Iron and Steel Institute*, 166: 213.

Richardson, H.K. (1935). Small cast thorium oxide crucibles. *Journal of the American Ceramic Society*, 18(1–12): 65–69.

Richerson, D.W. and Lee, W.E. (2018). *Modern ceramic engineering: Properties, processing, and use in design*. CRC Press, Boca Raton, FL.

Rigby, A. (2005). Controlling the processing parameters affecting the refractory requirements for Peirce-Smith converters and anode refining vessels. In: *Converter and fire refining practices, The Minerals Metals and Materials Society*, Warrendale, PA, pp. 213–222.

Rostoker, W. (1958). *The metallurgy of vanadium*. John Wiley, New York.

Routschka, G. (Ed.). (2001). *Refractory materials: pocket manual* (3rd Ed.). Vulkan-Verlag, Essen.

Routschka, G. and Granitzki, K.-E. (2005). Refractory ceramics. In: *Ullmann's encyclopedia of industrial chemistry*. Wiley-VCH Verlag GmbH and Co. KGaA, Weinheim.

Ruh, R. (1963). Reactions of zirconia and titanium at elevated temperatures. *Journal of the American Ceramic Society*, 46(7): 301–307.

Ruh, R. and Garrett, H.J. (1964). Reactions of zirconia and chromium. *Journal of the American Ceramic Society*, 47(12): 627–629.

Sadrnezhad, S.K. and Raz, S.B. (2005). Interaction between refractory crucible materials and the melted NiTi shape-memory alloy. *Metallurgical and Materials Transactions B*, 36(3): 395–403.

Saha, R.L. and Jacob, K.T. (1986). Casting of titanium and its alloys. *Defence Science Journal*, 36(2): 121–141.

Saha, R.L., Nandy, T.K., Misra, R.D.K., and Jacob, K.T. (1990). On the evaluation of stability of rare earth oxides as face coats for investment casting of titanium. *Metallurgical Transactions B*, 21(3): 559–566.

Sakamoto, K., Yoshikawa, K., Kusamichi, T., and Onoye, T. (1992). Changes in oxygen contents of titanium aluminides by vacuum induction, cold crucible induction and electron beam melting. *ISIJ International*, 32(5): 616–624.

Samsonov, G.V., Yasinskaya, G.A., and Wei, T.S. (1960). High-melting carbide, boride and nitride crucibles. *Refractories*, 1(1–2): 26–29.

Savitskii, E.M. and Burkhanov, G.S. (1970). Physical metallurgy of refractory metals and alloys, Consultants Bureau, New York.

Schafföner, S. (2020). Reactions of alkaline earth zirconate refractories with titanium alloys. In: *MATEC web of conferences* (Vol. 321). EDP Sciences, p. 10012.

Schafföner, S., Aneziris, C.G., Berek, H., Hubálková, J., and Priese, A. (2013). Fused calcium zirconate for refractory applications. *Journal of the European Ceramic Society*, 33(15–16): 3411–3418.

Schafföner, S., Aneziris, C.G., Berek, H., Hubálková, J., Rotmann, B., and Friedrich, B. (2015a). Corrosion behavior of calcium zirconate refractories in contact with titanium aluminide melts. *Journal of the European Ceramic Society*, 35(3): 1097–1106.

Schafföner, S., Aneziris, C.G., Berek, H., Rotmann, B., and Friedrich, B. (2015b). Investigating the corrosion resistance of calcium zirconate in contact with titanium alloy melts. *Journal of the European Ceramic Society*, 35(1): 259–266.

Schafföner, S., Qin, T., Fruhstorfer, J., Jahn, C., Schmidt, G., Jansen, H., and Aneziris, C.G. (2018). Refractory castables for titanium metallurgy based on calcium zirconate. *Materials and Design*, 148: 78–86.

Schlesinger, M.E. (1996). Refractories for copper production. *Mineral Processing and Extractive Metallurgy Review: An International Journal*, 16(2): 125–146.

Schlesinger, M.E. (2006). *Aluminum recycling*. CRC Press, Boca Raton, FL.

Schmutzler, H.J. and Sandhage, K.H. (1995). Transformation of Ba-Al-Si precursors to celsian by high-temperature oxidation and annealing. *Metallurgical and Materials Transactions B*, 26(1): 135–148.

Schnabel, M., Buhr, A., Buchel, G., Kockegey-Lorenz, R., and Dutton, J. (2011). Advantages of calcium hexaluminate in a corrosive environment. *Refractories World Forum*, 3(4): 87–94.

Schuster, J.C. and Palm, M. (2006). Reassessment of the binary aluminum-titanium phase diagram, *J. Phase Equilibria Diffusion*, 27(3): 255–277.

Schuyler, D.R., Petrusha, J.A., Hall, G.S., and Seagle, S.R. (1976). *Development of titanium alloy casting technology (No. 76–311721)*. AiResearch Manufacturing Company of Arizona, Phoenix, AZ.

Schwartz, M.A., White, G.D., and Curtis, C.E. (1953). *Crucible handbook a compilation of data on crucibles used for calcining, sintering, melting, and casting, united states atomic energy commission*. ORNL-1354, Oak Ridge National Laboratory, Oak Ridge, TN.

Seybolt, A.U. (1946). A practical furnace for vacuum melting. *Metal Progress*, 50: 1102.

Shedd, E.S., Marchant, J.D., and Henrie, T.A. (1964). *Continuous electrowinning of cerium metal from cerium oxides, report of Investigations 6362*. Bureau of Mines, U.S. Department of the Interior, Washington, DC.

Shunk, F.A. (1969). *Constitution of binary alloys*. Second Supplement, McGraw Hill, New York.

Siljan, O.-J., Schoning, C., and Grande, T. (2002). State-of-the-art alumino-silicate refractories for al electrolysis cells. *JOM: The Journal of The Minerals, Metals & Materials Society*, 54(5): 46–55.

Spedding, F.H. and Daane, A.H. (1952). The preparation of rare earth metals. *Journal of the American Chemical Society*, 74(11): 2783–2785.

Spedding, F.H. and Daane, A.H. (Eds.). (1961). *The rare earths*. Wiley, New York.

Spedding, F.H., Daane, A.H., Wakefield, G., and Dennison, D.H. (1960). Preparation and properties of high purity scandium metal. *Transactions of the Metallurgical Society of AIME*, 218: 608–611.

Spedding, F.H., Beaudry, B.J., Croat, J.J., and Palmer, P.E. (1968). The properties, preparation and handling of pure rare earth metals. In *Materials Technology—An Interamerican Approach*, p. 151, Am. Soc. Mech. Eng., New York.

Spedding, F.H., Beaudry, B.J., Croat, J.J., and Palmer, P.E. (1970). The preparation and properties of ultrapure metals. In: *Les Elements Des Terres Rares* (vol. 1, p. 25). Centre National de la Recherche Scientifique, Paris. Also R&D Report IS-2283, Ames Laboratory, USAEC, May.

Sundaram, C.V., Taneja, A.K., and Sridhar Rao, C.H. (1992). Technology trends in the extractive metallurgy of zirconium, titanium, tantalum and niobium. *Mineral Processing and Extractive Metallurgy Review*, 10(1): 239–265.

Taylor, L., Boyer, H. E. and Unterweiser, P. M. (Eds.) (1948). *Metals handbook*, American Society for Metals, Cleveland, OH.

Tetsui, T., Kobayashi, T., Mori, T., Kishimoto, T., and Harada, H. (2010). Evaluation of yttria applicability as a crucible for induction melting of TiAl alloy. *Materials Transactions*, 51(9): 1656–1662.

Tetsui, T., Kobayashi, T., Kishimoto, A., and Harada, H. (2012a). Structural optimization of an yttria crucible for melting TiAl alloy. *Intermetallics*, 20(1): 16–23.

Tetsui, T., Kobayashi, T., Kishimoto, A., and Harada, H. (2012b). Structural optimization of an yttria crucible for melting TiAl alloy. *Intermetallics*, 20(1): 16–23.

Theuerer, H.C. (1956). Removal of boron from silicon by hydrogen water vapor treatment. *JOM: The Journal of The Minerals, Metals & Materials Society*, 8(10): 1316–1319.

Thews, E.R. (1931). The suitability of various refractory materials for lead refining furnaces. *Feuerfest*, 7: 84.

Thompson, J.G. and Cleaves, H.E. (1939). Preparation of high-purity iron, RP1226. *Journal of Research of the National Bureau of Standards*, 23: 163–177.

Thornton, C.P. and Rehren, T. (2009). A truly refractory crucible from fourth millennium Tepe Hissar, Northeast Iran. *Journal of Archaeological Science*, 36(12): 2700–2712.

Tournier, C., Lorrain, B., Le Guyadec, F., Coudurier, L., and Eustathopoulos, N. (1998). Kinetics of interfacial reactions in molten U/solid Y_2O_3 system. *Journal of Nuclear Materials*, 254(2–3): 215–220.

Turner, D. (1931). Special refractories for metallurgical research. *Transactions of the Faraday Society*, 27: 112.

Tylecote, R.F. (1992). *A history of metallurgy* (2nd Ed.). Maney Publishing, London.

Uwanyuze, S., Kanyo, J., Myrick, S., and Schafföner, S. (2021). A review on alpha case formation and modeling of mass transfer during investment casting of titanium alloys. *Journal of Alloys and Compounds*, 865: 158558.

Van Humbeeck, J. (1999). Non-medical applications of shape memory alloys. *Materials Science and Engineering A*, 273: 134–148.

Vikulin, V.V., Rusin, M.Y., and Rusanova, L.N. (2006). Refractory materials and products based on natural wollastonite for making ceramic accessories to be used in aluminium industry. *Advances in Science and Technology*, 45: 2272–2277.

Wang, C.T., Baroch, E.F., Worcester, S.A., and Shen, Y.S. (1970). Preparation and properties of high-purity vanadium and V-15Cr-5Ti. *Metallurgical Transactions*, 1(6): 1683–1689.

Washburn, E.W. (1921). The dynamics of capillary flow. *Physical Review*, 17(3): 273.

Weber, B.C., Thompson, W.M., Bielstein, H.O., and Schwartz, M.A. (1957). Ceramic Crucible for Melting Titanium. *Journal of the American Ceramic Society*, 40(11): 363–373.

Weimer Alan, W. (Ed.). (1997). *Carbide, nitride and boride materials synthesis and processing*, Chapman and Hall, London.

Wilhelm, H.A. (1955). The preparation of uranium metal by the reduction of uranium tetrafluoride with magnesium. In: *Proceedings of the 1st international conference on the peaceful uses of atomic energy energy*. Geneva. Paper 817.

Wilhelm, H.A. (1960). Development of uranium metal production in America. *Journal of Chemical Education*, 37(2): 56–68.

Wilhelm, H.A. and Peterson, D.T. (1961). *Pure metals from chemical processes* (No. IS-394). Ames Lab, Ames, IA.

Wilhelm, H.A., Schmidt, F.A., and Ellis, T.G. (1966). Columbium metal by the aluminothermic reduction of Cb_2O_5. *JOM: The Journal of The Minerals, Metals & Materials Society*, 18(12): 1303–1308.

Wilson, R. (2000). *A practical approach to continuous casting of copper-based alloys and precious metals*. Institute of Materials, London.

Winkler, O.C. and Bakish, R.A. (Eds.). (1971). *Vacuum metallurgy*. Elsevier, Amsterdam.

Xiao, P. and Ralph, B. (Eds.). (2009). *Advanced ceramic materials, trans-tech, Zurich, Switzerland*.

Yazawa, A. and Kongoli, F. (2001). Liquidus surface of newly defined "ferrous calcium silicate slag" and its metallurgical implications. *High Temperature Materials and Processes*, 20(3–4): 201–208.

Yurkov, A. (2017). *Refractories for aluminum - electrolysis and the cast house*. Springer International Publishing, Cham, Switzerland.

Zhang, Z., Frenzel, J., Neuking, K., and Eggeler, G. (2005). On the reaction between NiTi melts and crucible graphite during vacuum induction melting of NiTi shape memory alloys. *Acta Materialia*, 53(14): 3971–3985.

Zhang, Z., Frenzel, J., Neuking, K., and Eggeler, G. (2006). Vacuum induction melting of ternary NiTiX (X= Cu, Fe, Hf, Zr) shape memory alloys using graphite crucibles. *Materials Transactions*, 47(3): 661–669.

Zhang, H.R., Tang, X.X., Zhou, L., Gao, M., Zhou, C.G., and Zhang, H. (2012). Interactions between Ni-44Ti-5Al-2Nb-Mo alloy and oxide ceramics during directional solidification process. *Journal of Materials Science*, 47(17): 6451–6458.

Zhang, H., Tang, X., Zhou, C., Zhang, H., and Zhang, S. (2013a). Comparison of directional solidification of γ-TiAl alloys in conventional Al2O3 and novel Y2O3-coated Al2O3 crucibles. *Journal of the European Ceramic Society*, 33(5): 925–934.

Zhang, Z., Zhu, K.L., Liu, L.J., Lu, X.G., Wu, G.X., and Li, C.H. (2013b). Preparation of BaZrO$_3$ crucible and its interfacial reaction with molten titanium alloys. *Journal of the Chinese Ceramic Society*, 41: 1272–1283.

Zhang, Y., Fang, Z.Z., Sun, P., Xia, Y., Lefler, H.D. and Zheng, S. (2020). Deoxygenation of titanium metal. In: Z.Z. Fang, F.H. Froes, and Y. Zhang (Eds.), *Extractive metallurgy of titanium*. Elsevier, Amsterdam, pp 181–223.

Zhao, L. and Sahajwalla, V. (2003). Interfacial phenomena during wetting of graphite/alumina mixtures by liquid iron. *ISIJ International*, 43(1): 1–6.

Index

For Product Safety Concerns and Information please contact our EU
representative GPSR@taylorandfrancis.com
Taylor & Francis Verlag GmbH, Kaufingerstraße 24, 80331 München, Germany

www.ingramcontent.com/pod-product-compliance
Lightning Source LLC
Chambersburg PA
CBHW060832170526
45158CB00001B/150